CARBONIFEROUS STRATIGRAPHY IN THE GRAND CANYON COUNTRY NORTHERN ARIZONA AND SOUTHERN NEVADA

Stanley S. Beus and Richard R. Rawson, ***Editors***
Department of Geology, Northern Arizona University
Flagstaff, Arizona

Field Trip No. 13
Ninth International Congress
of Carboniferous Stratigraphy and Geology

Published 1979 by
THE AMERICAN GEOLOGICAL INSTITUTE
as
AGI Selected Guidebook Series No. 2

Front Cover:

Marble Canyon, Arizona, from Dutton, C.E., 1882, *The Physical Geology of the Grand Canyon District*, U.S. Geological Survey Annual Report, Plate 35.

Inside Front Cover:

General Route Map

Back Cover:

Generalized Stratigraphic Column

Carboniferous Stratigraphy in the Grand Canyon Country, Northern Arizona and Southern Nevada
Guidebook for IX-ICC Field Trip No. 13
AGI Selected Guidebook Series No. 2

Printed in the United States of America

Library of Congress Catalog Card Number 78-74894
International Standard Book Number 0-913312-09-6
International Standard Serial Number 0190-6968

American Geological Institute
5205 Leesburg Pike
Falls Church, Virginia 22041

PREFACE

This guidebook was prepared for a field trip following the Ninth International Congress of Carboniferous Stratigraphy and Geology held at Urbana, Illinois, May 21–26, 1979. The field trip involved seven days and concentrated on the stratigraphy and paleontology of the upper Paleozoic, mainly Carboniferous, strata in the Grand Canyon and Mogollon Rim region of northern Arizona and the Spring Mountains and Arrow Canyon Ranges of southern Nevada.

The Carboniferous sections in Grand Canyon and southern Nevada are among the best known and most significant in western North America. The exposed strata formed in a variety of sedimentary environments reflecting a transition from the continental shelf to the Cordilleran geosyncline of the late Paleozoic. Numerous recent studies of Carboniferous lithostratigraphy and biostratigraphy in this region have added considerable detail and refinement to our understanding of the paleogeography of this region. This guidebook provides a summary of the present state of knowledge and, we hope, will suggest areas for further research.

We express appreciation to the researchers who have contributed papers to this guidebook and have assisted in the preparation of the road logs. Special thanks are extended to Edwin D. McKee, Ray C. Douglass, and Ivo Lucchitta of the U.S. Geological Survey and John J. Matthews of Union Oil Company, for consultation regarding many geologic details of the field trip. Financial aid in photographic documentation of the Nevada road logs was provided by the University of Illinois Graduate Research Board.

Stanley S. Beus
Richard R. Rawson
Editors

CONTENTS

ROAD LOG—FIRST DAY: VERDE VALLEY, ARIZONA

Ronald C. Blakey[1]

This field trip begins in Flagstaff, Arizona, and passes through Oak Creek Canyon, Sedona, the Verde Valley, Jerome, Perkinsville, Williams, and back to Flagstaff. It is a 140-mile loop that will afford a view of the entire Paleozoic section along the Mogollon Rim with specific stops to examine Devonian, Mississippian, and Pennsylvanian strata and fossils.

Mileage begins at Junction of Interstate Route 40 and U.S. Highway 89–A, at the south edge of Flagstaff.

Total Mileage

0.0 Junction of I–40 and U.S. 89–A; begin travel on 89–A south toward Sedona.

0.4

0.4 Junction with Lake Mary Road; low bluffs to west in University Heights Subdivision are Kaibab Limestone. Continue on U.S. 89–A.

1.1

1.5 Roadcut in Kaibab Limestone.

1.4

2.9 Junction with Airport Road; continue on U.S. 89–A. Tertiary basalt in roadcuts next several miles. Highway is traversing Coconino Plateau, one of the largest plateaus on the Colorado Plateau. Dissected upland surface is chiefly underlain by Kaibab Limestone. In this area an irregular veneer of basalt caps the plateau. Flows originated from fissures, vents, and cinder cones.

The Kaibab Limestone of Leonardian age was deposited in and adjacent to the Permian seaway. Fossils indicate that both restricted and normal marine conditions prevailed. Detailed depositional analysis is difficult because of intensive dolomitization. The eastern Permian shoreline lay some 40 to 50 miles (46 to 80 km) to the east.

4.0

6.9 Escarpment at 12 o'clock is upthrown block of Oak Creek Fault. This normal fault trends north-south from near Flagstaff to southeast of Sedona. Part of Flagstaff's water supply comes from wells associated with fractured Coconino Sandstone along fault near Woody Mountain. Major drainage system to east of road is Pump House Wash, a major tributary to Oak Creek.

5.0

11.9 **STOP 1—OAK CREEK CANYON VISTA**

0.3

12.2 Excellent roadcut in Tertiary basalt and red cinder zone. Well-developed columnar jointing present in many basalts in this area.

0.8

13.0 Oak Creek Fault Zone; Kaibab Limestone exposed on eastern downthrown side and Coconino Sandstone on upthrown western side.

0.1

13.1 Cross-stratified Coconino Sandstone; the Leonardian Coconino Sandstone is part of the "great sand pile" on the Colorado Plateau. Although desperately in need of a modern sedimentological study, the Coconino has long been considered a classic eolian deposit. Winds blowing from the north along the edge of the Permian Cordilleran Seaway moved large quantities of sand south across a low coastal plain. Numerous vertebrate trackways are preserved on the dune forests a few miles west of here. The Coconino is a very fine-grained, well-sorted quartz arenite.

0.1

13.2 Recross fault zone; roadcuts in Kaibab Limestone and talus and landslide deposits.

0.4

13.6 Approximate position of Kaibab-Coconino contact.

0.4

14.0 Switchback crosses then recrosses fault zone. Both sides in Coconino Sandstone.

0.3

14.3 Bridge over mouth of Pump House Wash. Road now follows course of Oak Creek. Oak Creek is a spring-fed tributary to Verde River. Springs are associated with faults and joints and generally discharge from the Coconino. A state fish hatchery uses the spring water to raise trout for stocking the creek.

0.6

14.9 Spectacular debris and landslide deposits on both sides of Oak Creek.

0.1

15.0 Pine Flat Campground.

1.0

16.0 Stratigraphy on west wall of canyon: red cross-stratified sandstone is Schnebly Hill formation (upper member of Supai Formation). Light-colored cross-stratified sandstone is Coconino. Transition between the two formations is well displayed.

The Schnebly Hill formation is informally introduced in this guidebook for a post-Supai, pre-Coconino sand body present throughout the Mogollon Rim but absent in the Grand Canyon. The unit was deposited as a complex sabkha, tidal, beach, and eolian system in the southwest flank of the Holbrook Basin (Blakey, this guidebook). A similar unit on the northeast flank is the De Chelly Sandstone.

0.5

16.5 The yellowish-orange flat-bedded sandstone marks a controversial unit present in the Coconino Sandstone.

[1] Department of Geology, Northern Arizona University, Flagstaff, Arizona 86011.

Early studies in the region suggested that the "green line" of concentrated vegetation growing at this horizon marked the position of an erosional unconformity separating the Coconino Sandstone from the overlying Toroweap Formation. However, sedimentological and stratigraphic studies have shown that it is not practical to separate the two units. Thus the entire sequence is called the Coconino Sandstone even though the unit above the "green line" is the Toroweap equivalent.

1.1

17.6 Mouth of West Fork of Oak Creek; this is a popular hiking area and provides excellent access to upper part of Permian sequence.

0.7

18.3 Don Hoel's Cabins.

0.7

19.0 Campground; large sets of cross-beds in Coconino exposed along road.

0.4

19.4 Junipine Cabins; several basalt dikes cut Coconino Sandstone on east side of highway. West wall of Oak Creek exposes Schnebly Hill, Coconino, and Kaibab.

0.9

20.3 Drag folding in Schnebly Hill associated with Oak Creek Faulting zone.

0.2

20.5 Slide Rock area. This famous area in Oak Creek attracts over 2000 people daily during the summer. The creek plunges through a narrow chute in the Schnebly Hill that is a popular "ride" for adventurous swimmers. Prominent limestone ledge on west wall is Fort Apache Limestone Member.

0.3

20.8 Oak Creek Bridge. Schnebly Hill exposed on both sides of road. Much of the unit is bleached on east side of canyon; this is probably due to water migration through fractured rocks associated with fault zone.

1.0

21.8 Manzanita Campground.

0.4

22.2 Red sandstone in roadcuts is Schnebly Hill formation.

0.7

22.9 Encinosa picnic area. Large-scale cross-stratification developed in Schnebly Hill.

1.0

23.9 Indian Gardens; this was a popular gathering place for Indians in the region. Evidence of farming is present along the flood plain of Oak Creek. Munds Canyon enters from east.

1.1

25.0 Fault zone with reverse drag. Esplanade Sandstone and Hermit Shale exposed in upturned sequence. Roadcuts display spectacular channels in Hermit; Esplanade forms inner walls of canyon. These two terms are introduced into the Mogollon Rim from the Grand Canyon. Both units can be correlated from the western Grand Canyon into the Mogollon Rim. The Hermit was deposited on a low-lying plain by fluvial and possibly tidal processes. The Esplanade represents shoreline and shallow marine deposition. The Hermit is late Wolfcampian to Leonardian and the Esplanade is Wolfcampian.

0.7

25.7 Grasshopper Point. Excellent exposures of Permian section from Esplanade to Coconino.

0.5

26.2 **STOP 2—MIDGLEY BRIDGE.**

0.7

26.9 Entering Sedona (sign); contact between Esplanade and Hermit; road descends into Esplanade Sandstone.

0.7

27.6 Cross Sedona Fault; enter downtown Sedona; bedrock is Hermit Shale.

0.6

28.2 Junction with Arizona Highway 179 (the "Y"); bear right on U.S. 89–A. Low, juniper-covered hills composed of Hermit Shale. Excellent roadcuts displaying bedding for next mile.

0.6

28.8 Spectacular channels in Hermit Shale exposed in south roadcut. At 2 o'clock is Coffee Pot Rock, a famous Sedona landmark in Schnebly Hill formation.

0.9

29.7 Sedona airport to south is on top of lava-capped mesa.

0.6

30.3 Gray Mountain (Capitol Butte) at 3 o'clock (due north) exposes Schnebly Hill and Coconino. Note "green line" near top.

0.9

31.2 View of Mogollon Rim and upper Dry Creek Valley to north.

0.4

31.6 Leaving Sedona; roadcuts are in Hermit. Submarine rock is at 3 o'clock.

0.8

32.4 Junction with Red Rock Loop Road East. This road leads to Cathedral Rock and Red Rock Crossing, one of the most famous views in the west. Continue on U.S. 89–A.

0.5

32.9 Road crosses Cathedral Rock Fault. North or upthrown block approximately 700 to 1000 feet (215–305 m) stratigraphically higher than south block. Cross-stratified, pale grayish-orange, fractured sandstone is Esplanade Sandstone, the top of the Supai Group. Hermit and Schnebly Hill exposed on downthrown side. Basalt-capped butte to south is Scheurman Mountain. View to west clearly shows stratigraphic displacement on fault.

0.7

33.6 Red Rock Loop Road West. At 3 o'clock (due north) Dry Creek has incised into upthrown fault block and entire Supai Group and top of Redwall Limestone are exposed. Continue on U.S. 89–A. Slow down for STOP 3 ahead.

0.3

33.9 **STOP 3—ENTRENCHED MEANDER ON DRY CREEK.** Walk (or drive if size of vehicle will allow) northward through cattle gate about ¼ mile to sandstone cliffs of Dry Creek exposures. Return to vehicle and continue on U.S. 89–A.

1.2

35.1 Dry Creek on north side of road; roadcuts in Hermit.

0.5

35.6 Bridge across Dry Creek. Sandstone at creek level is probably top of Esplanade Sandstone.

0.4

36.0 Fault; Schnebly Hill or Hermit on east side, Tertiary Hickey basalts and gravels on west side. Low hills to west composed of Tertiary Verde Formation.

0.9

36.9 Black Mountain at mouth of Sycamore Canyon at 3 o'clock. Excellent view of Verde Valley and Mingus Mountain straight ahead.

1.8

38.7 Roadcut in basalt and cinders of mid-Tertiary age.

0.8

39.5 Junction with Page Springs Road. Continue on U.S. 89–A.

0.8

40.3 Bridge over Spring Creek; outcrops of Tertiary Verde Formation to north. Redbed facies of Verde is reworked Permian sediments. Thin lacustrine limestones of the Verde cap mesas.

1.3

41.6 Reddish-orange clastics in roadcuts probably in Verde Formation. Broad, flat-topped mountain at 7 o'clock (southeast) is House Mountain; some class this late Tertiary volcanic vent as a shield volcano.

2.1

43.7 Excellent views of south side of Verde Valley. Note pediment surfaces. Town of Jerome on side of Mingus Mountain beneath block letter "J" at 2 o'clock (west). Sprawling town in Verde Valley is Cottonwood.

1.3

45.0 Junction with Cornville Road. Continue on U.S. 89–A.

0.4

45.4 Limestone in roadcut is Verde Formation.

0.7

46.1 Bridge across Verde River; enter Cottonwood.

0.8

46.9 Merge with Arizona 279. Bear right on U.S. 89–A.

0.6

47.5 Stoplight; turn left on U.S. 89–A south toward Jerome. At 12 to 3 o'clock is western clastic facies of Verde Formation. This alluvial and lacustrine unit grades eastward into limestone and mudstone of eastern facies of Verde. During the late Cenozoic the course of the Verde River was blocked and dammed by subsidence and/or lava flows. A large lake was formed and in the center of the lake fossiliferous fine-grained clastics and carbonates were deposited. Coarse clastics rim the basin and intertongue with finer deposits. Several lava flows interbedded within sequences.

0.3

47.8 Gravel pit operations in terrace gravels of the Verde River. Much evidence of the past mining history of the region can be seen in the decaying structures.

1.3

49.1 Junction with Cottonwood Airport Road. Continue on U.S. 89–A.

2.4

51.5 Junction; stop sign; turn left to Jerome. The Phoenix Cement Company (on right, west) uses limestone and clay from Redwall Limestone and Verde Formation to make cement.

1.0

52.5 Begin ascent to Jerome through Verde Formation gravels.

0.7

53.2 Martin Formation (Devonian) and Redwall Limestone exposed in roadcuts for next several miles. Numerous small faults in area.

0.6

53.8 Laminated aphanitic dolomite of Martin Formation exposed in roadcut. The Martin Formation of Late Devonian age (Frasnian) is dolomite, sandy dolomite, and dolomitic siltstone deposited in diverse shoreline, tidal, and supratidal environments.

0.5

54.3 **STOP 4—DEVONIAN FOSSIL LOCALITY IN MARTIN FORMATION, DECEPTION GULCH.**

0.3

54.6 Basalt roadcut; numerous dikes and altered zones.

0.8

55.4 Old Mingus Union High School; enter Jerome.

0.1

55.5 Junction with road to State Historical Museum; continue straight ahead.

0.2

55.7 Badly deformed sidewalk marks position of Verde Fault. Jerome straddles this large normal fault and is slowly sliding down slope due to extensive creep. Note retaining walls which have attempted unsuccessfully to slow this process.

0.5

56.2 Junction with road to Perkinsville; bear right on gravel road. Road passes near open pit of United Verde Mine.

0.7

56.9 Precambrian schists of "Yavapai Group."

0.3

57.2 Recross Verde Fault into Paleozoic sedimentary rocks. Road climbs through section in Martin Formation.

1.4

58.9 Roadcut in Martin-Redwall contact. Red butte at 12 o'clock (southwest) is Supai Group.

0.1

58.7 Numerous faults; road passes through Redwall, Supai, and then Martin. Excellent exposures of Martin at 59.0 miles.

0.4

59.1 One-lane bridge; Martin Formation.

0.1

59.2 Supai Group straight ahead. Capping unit is Miocene Hickey Basalt. Red sandstone in roadcut is probably Manakacha Formation.

0.6

59.8 Supai-Redwall contact. Deeply weathered Redwall grades upward into limey mudstone of Watahomigi Formation. No basal conglomerate present here. Road continues along this horizon.

0.5

60.3 Outcrops of Watahomigi Formation.

0.1

60.4 Manakacha cliff. Manakacha cliff unit can be traced throughout Verde and Chino Valleys and westward into Grand Canyon. This is a key unit in the stratigraphy of the Supai Group.

0.1

60.5 **STOP 5—FIRST VIEW AREA.** (Note: several brief stops will be made, time permitting, over the next several miles where Redwall Limestone and Watahomigi, Manakacha and Wescogame Formations can be examined.)

0.3

60.8 Outcrops of Wescogame Formation.

0.1

60.9 "First View"; basalt caps low hills.

0.2

61.1 Wescogame Formation.

0.3

61.4 Manakacha cliff; Watahomigi exposed in washes below road level.

0.6

62.0 Possible Manakacha-Wescogame boundary. This is a difficult horizon to locate in western and central Mogollon Rim.

0.4

62.4 Deep roadcut into top of Manakacha cliff. Excellent view to west reveals following features from left to right: Juniper Mountains, Chino Valley, Big Black Mesa, Picacho Butte, Mount Floyd, and Mogollon Rim. Low, lava-capped butte in foreground has excellent exposures of upper Redwall, Watahomigi, Manakacha, and Wescogame. Road descends through Manakacha Formation.

0.3

62.7 Watahomigi Formation.

0.2

62.9 Mudstone and limestone of Watahomigi.

0.1

63.0 Fault; Manakacha cliff; road descends into lower Manakacha.

0.3

63.3 Badly weathered carbonate may be Watahomigi or Redwall.

0.1

63.4 Watahomigi mudstone.

0.3

63.7 Manakacha cliff.

0.2

63.9 Watahomigi mudstone and carbonate.

0.3

64.2 Hill referred to at 62.4 miles at right (north) of road. Possible stop.

0.1

64.3 Junction with Forest Service Road 155. Continue on Perkinsville road.

0.1

64.4 Fault; Manakacha cliff.

0.2

64.6 Excellent northward view to Perkinsville and upper Verde Valley and Mogollon Rim.

0.5

65.1 Junction; turn right (north); Redwall and Watahomigi compose small butte on north side of junction. Road continues on pediment surface carved on Redwall Limestone.

1.6

66.7 Mudstone and carbonate of Watahomigi Formation. Excellent view to north of nearly complete Redwall Limestone. Ledgy dolomite of Martin Formation at base of section. Road follows Orchard Draw.

0.6

67.3 Excellent view of the mouth of Sycamore Canyon directly below the San Francisco Peaks on the skyline.

2.7

70.0 Note faulted and duplicated Supai Group straight ahead. Redwall and lower Supai Group exposed to east of road. Note faulting.

0.7

70.7 Basalt in roadcut.

1.0

71.7 Junction with Chino Valley Road; continue straight ahead.

0.7

72.4 Nearly entire Redwall and Supai exposed in cliffs straight ahead.

0.9

73.3 Junction with north Chino Valley Road; continue straight ahead.

0.4

73.7 Perkinsville Ranch; keep to left.

0.3

74.0 Bridge over Verde River; cliffs to southwest on river are sandstones in Manakacha Formation.

0.1

74.1 Turn off right to **STOP 6—PERKINSVILLE.** Proceed to abandoned Perkinsville RR station. Retrace steps to road.

0.3

74.4 Cross railroad tracks. Turn off left on dirt road to **STOP 7—WEST PERKINSVILLE.** Return to main road.

0.1

74.5 Manakacha cliff to west of road. Road begins descent out of Verde Valley through basalt, gravel, and silt of Perkinsville Formation. Entire Carboniferous section exposed in cliffs to east. Esplanade Sandstone caps the escarpment.

1.8

76.3 View of Mogollon Rim; Coconino, Toroweap, and Kaibab exposed.

0.9

77.2 Junction; continue straight.

1.5

78.7 Government Canyon.

0.9

79.6 Junction; turn left (west); prominent "white stripe" in Mogollon Rim is limestone in Toroweap Formation which pinches out west of Sycamore Canyon. East of Sycamore Canyon the Toroweap grades into Coconino sandstones.

1.1

80.7 Junction with Bear Siding Road; continue straight. Note: Excellent outcrops of Redwall and Supai exposed along Verde River five miles to south on this road. An abandoned low-grade copper mine, the U.S. Mine, is at end of this road.

0.4

81.1 Bear Canyon; cross-stratified sandstone in canyon is Esplanade Sandstone; the Hermit Shale forms the two buttes to southwest and underlies the low area to north.

1.6

82.7 Pavement begins.

0.3

83.0 Junction with Drake Road; continue right on paved highway.

2.7

85.7 Upper Bear Canyon to east of road; Coconino Sandstone exposed in bottom of canyon.

3.0

88.7 Enter Kaibab National Forest. Road is climbing lava ramp on Mogollon Rim. Basalt, cinders, and related units cap this portion of the Coconino Plateau.

14.9

103.6 Bill Williams Mountain on left. This complex Tertiary volcanic feature is the oldest volcano in the San Francisco Volcanic Field.

2.4

106.0 Road enters canyon cut into Tertiary Basalt.

1.0

107.0 Reservoir for City of Williams water supply.

0.4

107.4 Enter Williams.

0.3

107.7 Junction with U.S. 66. Turn right (east) toward Flagstaff. Begin Arizona milepost system on U.S. 66 with MP 163.5.

1.4

109.1 Overpass over Sante Fe Railroad; leave Williams. MP 164.9.

1.1

110.2 Junction with Arizona 164 to Grand Canyon on left. MP 166.0

1.2

111.4 Garland Prairie exit off I-40; San Francisco Peaks at 11 o'clock. MP 167.2.

2.8

114.2 Intermittent lake (open park land) on both sides of road. MP 170.0.

0.8

115.0 Roadcut in red cinders. MP 170.8.

0.5

115.5 Deer farm road on left; Ponderosa Motel on right. MP 171.3.

1.7

117.2 Roadcut in red cinders and volcanic ash. MP 173.0.

1.2

118.4 Roadcut in basalt. MP 174.2.

0.9

119.3 Roadcut in cinders and volcanic ash. MP 175.1.

0.8

120.1 Spitz Spring road exit. MP 175.9.

2.2

122.3 Parks road exit. MP 178.1.

2.6

124.9 Roadcut in basalt. MP 180.7.

1.1

126.0 Highway rest stop on right. MP 181.8.

0.4

126.4 Partially quarried cinder cone on right. MP 182.2.

1.8

128.2 Navajo Army Depot on right; now closed except for maintenance crews. During World War II and the Korean War this was a major storage site for ammunition and explosives. MP 184.0.

3.1

131.3 Naval Observatory on skyline at 12 o'clock. MP 187.1.

1.4

132.7 Roadcut in basalt on left. MP 188.5.

1.5

134.2 A–1 Mountain exit. MP 190.0.

1.2

135.4 First Flagstaff exit; roadcuts in basalt and cinders; stay on I–40 east. MP 191.2.

1.3

136.7 San Francisco Peaks and Mt. Elden on left (north). These major volcanic features will be discussed in more detail on second day field trip. MP 192.5.

0.5

137.2 Flagstaff city limits. MP 193.0.

2.0

139.2 University Heights on right. MP 195.0.

0.4

139.6 Sedona-Phoenix exit on right; stay on I–40 to next exit. MP 195.4.

0.2

139.8 Flagstaff exit; turn right. Roadcut in upper Kaibab Limestone; abundant molds of Permian fossil mollusks, brachiopods, and trilobites occur in this part of the Kaibab. Follow U.S. 89 north 2 miles to NAU campus.

END OF FIRST DAY ROAD LOG

STOP DESCRIPTIONS—FIRST DAY

Ronald C. Blakey[1]

STOP 1—OAK CREEK CANYON VISTA
Sec. 14, T. 19 N., R. 6 E.
Coconino County, Arizona

Oak Creek is a southward-flowing permanent stream that has cut a significant canyon reentrant into the Mogollon Rim. Pump House Wash, the major tributary, begins some 8 miles (11 km) north of the view point and can be seen to the east. To the south is Oak Creek Canyon and to the west is Sterling Canyon and the five switchbacks that lower U.S. Highway 89–A into the canyon.

The course of the canyon is controlled by late Tertiary faulting. The main Oak Creek Fault is several hundred meters west of the vista (fig. 1). The upthrown west side forms the escarpment that the highway has been following from Flagstaff. The obviously higher west wall of Oak Creek Canyon is due to the vertical throw of the fault. Off-set Paleozoic sedimentary rocks and Tertiary basalts document the 700– to 1500–foot (210–450 m) displacement.

The oldest rocks visible from the vista are the reddish-orange sandstone of the "upper Supai Formation," referred to informally as the Schnebly Hill formation in this guidebook. This is succeeded by cliff-forming grayish-orange sandstone of the Coconino Sandstone. A ledge is present near the upper third of the Coconino that marks the position of the Toroweap-equivalent portion of the section. Although the Toroweap Formation has previously been recognized in Oak Creek Canyon, sedimentological and stratigraphic studies have demonstrated the impracticality of separating the two units in this area. The Coconino is overlain by yellowish-gray dolomite, dolomitic limestone, and dolomitic sandstone of the Kaibab Limestone. The Kaibab is the youngest Paleozoic formation on the Mogollon Rim.

Tertiary olivine basalts cap parts of the Mogollon Rim in this area and have been dated at 6 to 8 million years in the Oak Creek area (McKee and McKee, 1972). The basalts originated from fissures, cones, and vents and are interbedded with cinder and ash deposits.

STOP 2—MIDGLEY BRIDGE
Sec 4, T. 17 N., R. 6 E.
Coconino County, Arizona

Midgley Bridge is near the southern end of Oak Creek Canyon where the canyon has widened considerably. The canyon walls in this area expose the entire Permian and a slight portion of the Pennsylvanian section. Here the old three-fold division of the "Supai Formation" (McKee, 1945) is easily seen. The inner canyon of Oak Creek exposes the lower or "C" member; the red, juniper-covered slopes are composed of the middle or "B" member; the towering reddish-orange cliffs are composed of the upper or "A" member (fig. 2).

In the Grand Canyon, careful stratigraphic and paleontologic work by McKee (1975) and others has rather precisely located the position of the Pennsylvanian-Permian boundary. Careful correlation from the western Grand Canyon to Chino Point and then eastward through Chino and Verde Valleys into the Sedona area has demonstrated that this boundary is present near creek level below Midgley Bridge. The unconformity can be traced more than a mile along the creek. At places only a sharp bedding plane marks the break; other places tens of feet (up to 10 m) of relief can be observed. About 100 yards (m) southwest of the bridge near creek level an irregular surface is exposed. This surface is locally overlain by limestone- and siltstone-pebble conglomerate. At least 30 feet (10 m) of relief can be seen in a cliff wall immediately above the irrigation ditch on the west side of Oak Creek. The underlying unit, the Wescogame Formation of Pennsylvanian age (Virgilian), consists of dolomitic silty sandstone. Maximum amount exposed is less than 40 feet (13 m). The overlying Esplanade Sandstone of Wolfcampian age comprises dolomitic silty sandstone, fine-grained quartz arenite, and irregular mudstone. The bulk of the unit displays medium- to large-scale tabular-planar to wedge-trough cross stratification.

Because the Esplanade Sandstone (which forms the bulk of the "lower member of the Supai Formation") is the top formation in the Supai Group, all the overlying units of the "Supai Formation" in Oak Creek Canyon do not correlate with the type Supai. The following units are herein recognized: 1) The slope-forming unit of red sandstone, siltstone, limestone, and conglomerate is correlated with the Hermit Shale. The term Hermit is introduced into the Mogollon Rim. This is the so-called "middle member of the Supai Formation." 2) The cliff-forming reddish-orange sandstone with the Fort Apache Limestone marker near the middle is a stratigraphic sequence not recognized in the Grand Canyon. It is informally herein called the Schnebly Hill formation. It grades upward into and intertongues with the Coconino Sandstone.

STOP 3—ENTRENCHED MEANDER ON DRY CREEK
Secs. 16, 17, T. 17 N., R. 5 E.
Yavapai County, Arizona

Dry Creek forms a major drainage between Oak Creek and Sycamore Canyons. Where the dry wash crosses the upthrown block of the Cathedral Rock Fault, the entire Supai Group and top of the Redwall Limestone are exposed.

The top 10 to 20 feet (3 to 6 m) of the Redwall are exposed in the creek bottom. The Redwall comprises grayish-white, crystalline, crinoidal calcarenite. A few spiriferid brachiopods and rugose corals are also present. Although not well exposed, the top of the Redwall does not appear to have a karst surface. Rather, a planar surface marks the contact.

All four formations of the Supai Group are exposed. The Watahomigi Formation consists of approximately 65 feet (20 m) of reddish-brown siltstone, dolomite, and dolomitic sandstone. The unit forms slopes and weak ledges. An erosion surface with two feet of local relief separates the overlying Manakacha Formation. Thirty-five feet of cross-stratified sandstone and dolomitic siltstone form the bulk of this unit. It

---

[1] Department of Geology, Northern Arizona University, Flagstaff, Arizona 86001.

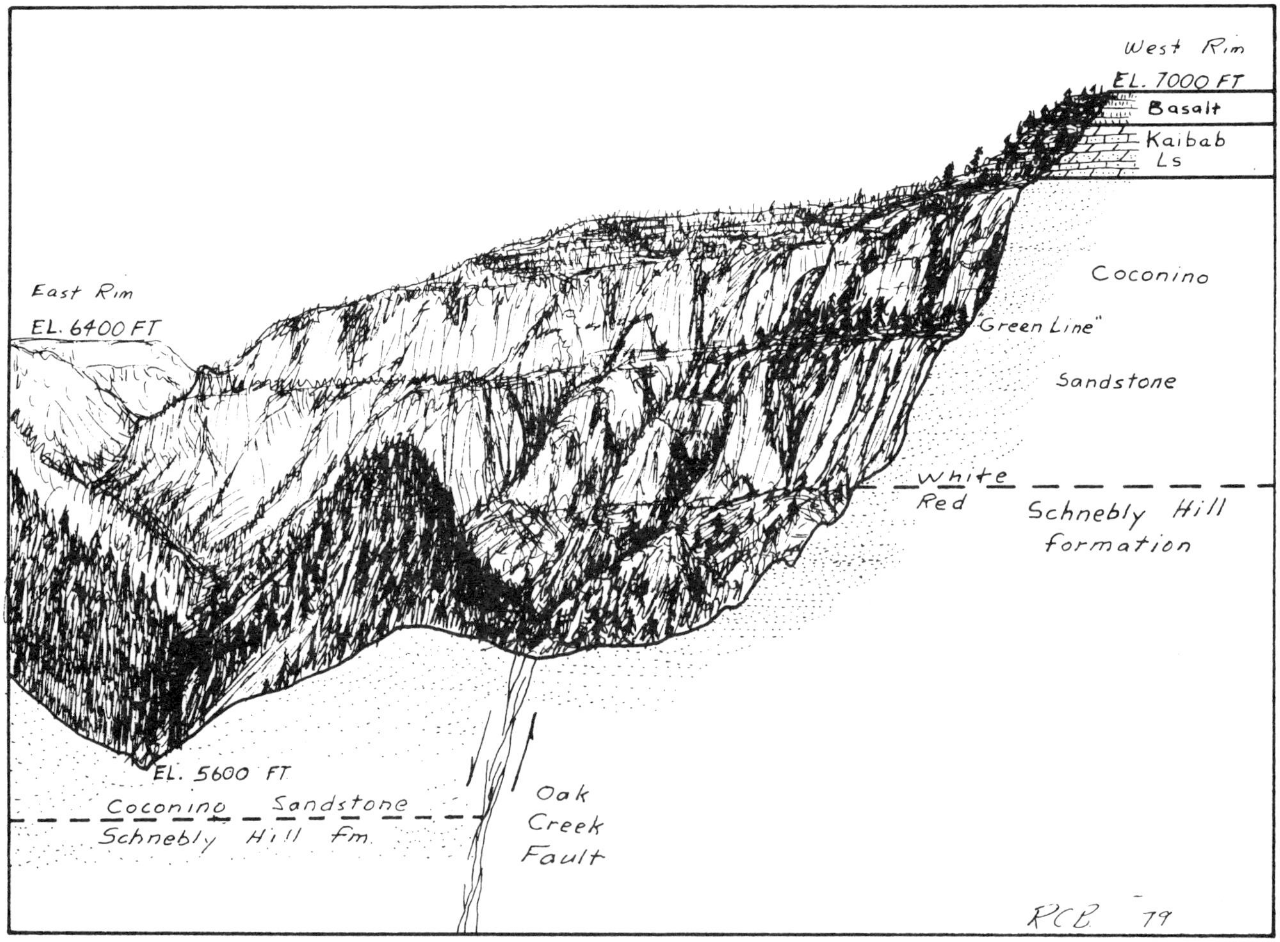

Figure 1. Sketch of view looking south from Oak Creek Canyon vista. Tertiary basalt caps the east rim (elev. 6,400 ft (2,020 m)), the conical hill in the left central portion of the sketch, and widely scattered areas of the west rim.

forms a reddish-brown cliff-forming marker bed. The Wescogame Formation is reddish-brown to grayish-pink cross-stratified sandstone and poorly exposed siltstone. The 145-foot-thick sequence forms cliffs and steep slopes. The Esplanade Sandstone overlies the Wescogame and the boundary marks the Pennsylvanian-Permian boundary. The contact is not exposed but is marked by a 10- to 20-foot slope. At least 120 feet (38 m) are present but the section is incomplete owing to removal by faulting and erosion.

The Dry Creek section exposes the only complete Pennsylvanian section between Sycamore Canyon and Fossil Creek. The latter exposures, some 40 miles to the southeast, drastically differ from the Dry Creek section. The Naco Group at Fossil Creek is related to deposition in the Pedregosa Basin whereas the Supai Group is related to deposition on the shelf bordering the Cordilleran Miogeosyncline. The Sedona area is near the crest of a large low arch that separates the two depositional systems.

STOP 4—DEVONIAN FOSSIL LOCALITY IN MARTIN FORMATION, DECEPTION GULCH
SE ¼ sec. 23, T. 16 N., R. 2 E.

One of the richest Devonian fossil localities in the Verde Valley occurs here in the upper part of the Jerome Member of the Martin Formation (Beus, 1978). The fossiliferous beds crop out as 0.5-1.0 m thick ledges of pinkish-gray dolomite about 100 m north of the road (U.S. 89-A), east of the road cut and about 12 m stratigraphically above the road in a small ravine. The exposures are on the downthrown side of the main Verde Fault on the southeast edge of Jerome. Although the Martin-Redwall contact (Devonian-Mississippian boundary) is not exposed here the fossil beds are believed to be within 30 m of the top of the Devonian. The fossils are partially to completely silicified and include the brachiopods *Atribonium arizonensis, Cupularostrum saxatilis, Cyrtospirifer whitneyi, Indospirifer orestes, Schuchertella prava,* and *Tenticospirifer cyrtinaformis*; the corals *Aulopora* sp., *Disphyllum* sp., *Hexagonaria* sp., and *Thamnopora* sp.; and *Amphipora* sp. and tabular forms of stromatoporoids.

STOP 5—FIRST VIEW AREA
T. 17 N., R. 2 E.
Yavapai County, Arizona

The northern slopes of Mingus and Woodchute Mountains have abundant exposures of the Redwall Limestone and Supai Group. Because of extensive faulting, complete sections are not exposed; however, numerous partial sections can be pieced together and the resulting section ranges from Martin to middle Wescogame.

Two general lithologic types compose the Redwall Limestone in the Verde Valley. Dominant is light-gray, crystalline, crinoidal calcarenite that forms massive cliffs. At places this is overlain by dense, yellowish- to purplish-gray, recrystallized micritic limestone. The latter unit is not always present. In the First View area the crinoidal calcarenite composes most of the Redwall. The upper portion is deeply weathered and the upper unit is absent or so deeply weathered as to be unrecognizable.

The Watahomigi Formation is 100 to 150 feet (30-50 m) thick and comprises three general subdivisions. Basal purplish-gray limey mudstone is succeeded by grayish ledge-forming recrystallized micrite which is overlain by red silty sandstone and siltstone. Bioturbation is abundant in much of the upper part of the formation.

The Manakacha Formation is approximately 75 feet (25 m) thick and is also divisible into three units. The lowest unit is identical to the upper Watahomigi. Near Perkinsville a conglomerate separates the two units but where this is absent a limey concretionary zone is thought to mark the contact. The diagnostic portion of the Manakacha is cliff-forming limey sandstone. This Manakacha cliff unit forms a key stratigraphic horizon in the Supai Group in the Mogollon Rim area. It consists of low-angle tabular-planar or wedge-planar cross-stratification with southerly dipping laminae. Limey concretions are abundant in some units. At First View the cliff unit has two cross-stratified sequences separated by red siltstone. Bioturbation is obvious in much of the cliff sequences. The upper Manakacha comprises siltstone and limey silty sandstone. The contact with the overlying Wescogame is difficult to pick and is generally placed at the base of the interval where cyclic ledge- and slope-forming cross-stratified sandstone and limey siltstone dominate the section.

Only part of the Wescogame Formation is exposed in the First View area. Better and more complete sections are exposed in the Perkinsville area. Cross-stratification is generally of a smaller scale and exhibits higher angles of dip than the Manakacha Formation.

STOP 6—PERKINSVILLE
Sec. 32, T. 18 N., R. 2 E.
Yavapai County, Arizona

The cliffs northeast of the Perkinsville railroad station expose a complete Carboniferous section. Faulting repeats the section a few miles to the east and the section is also well exposed in the cliffs of the Verde River between Perkinsville and Sycamore Canyon.

The crinoidal calcarenite unit dominates the Redwall Limestone. An extensive karst surface overlies the Redwall and at Sycamore Canyon tens of feet of relief are exposed on the east side of the Canyon. This surface is not well exposed at Perkinsville.

Basal conglomerate of the Watahomigi Formation is present at both Perkinsville and Sycamore Canyon. It consists of angular, gravel-sized red, gray, brown, and yellow chert fragments and quartz sandstone. The chert was presumably derived from the underlying Redwall. The remainder of the Watahomigi and Manakacha Formations are similar to the First View section. The Watahomigi and Manakacha Formations are separated by well-developed conglomerate.

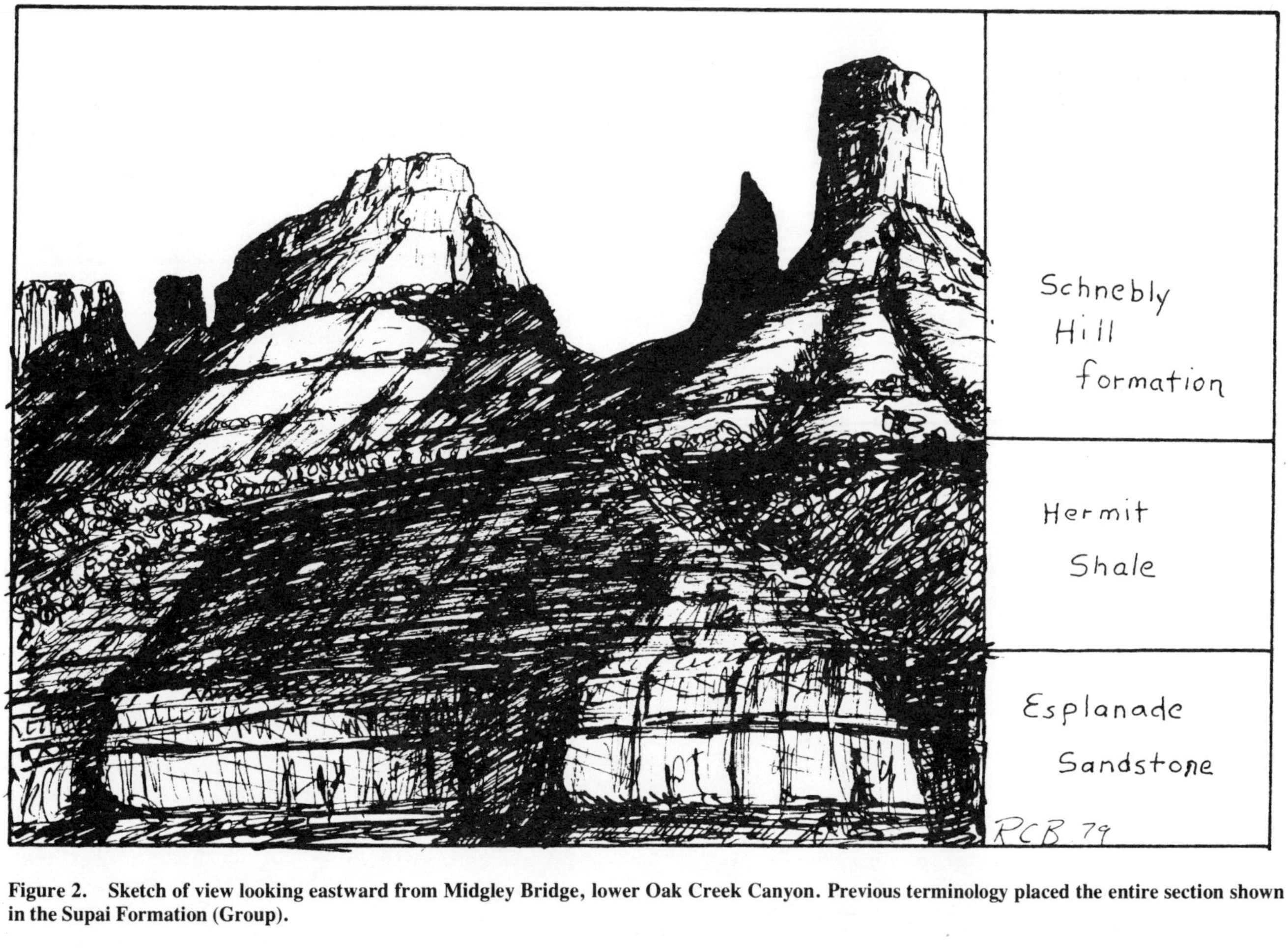

Figure 2. Sketch of view looking eastward from Midgley Bridge, lower Oak Creek Canyon. Previous terminology placed the entire section shown in the Supai Formation (Group).

The entire Wescogame Formation is exposed at Perkinsville. Approximately 180 feet (60 m) of cross-stratified sandstone, limey siltstone, bioturbated sandstone, and minor micritic limestone compose the ledge-, cliff-, and slope-forming sequence.

Overlying the Wescogame Formation with apparent erosional unconformity is the Esplanade Sandstone. Although conglomerate is absent at this locality, the thicker cross-stratified sets of the Esplanade contrast with the thinner ledges of the Wescogame. The contact is especially noticeable from a distance. The block "P" above Perkinsville is painted on the lower portion of the Esplanade.

STOP 7—WEST PERKINSVILLE
Sec. 31, T. 18 N., R. 2 E.
Yavapai County, Arizona

Throughout much of the Mogollon Rim the Supai Group weathers as a narrow, steep, broken escarpment. The straight, commonly talus-covered outcrop is difficult to approach with a three-dimensional study. An exception to this is an outcrop of the Redwall, Watahomigi, and Manakacha Formations exposed several hundred meters west of the Perkinsville Road along the Santa Fe Railroad tracks. Here the outcrop is exposed as broad benches and undercut ledges providing an excellent opportunity to study units that are generally not well exposed.

The general description of the units exposed is similar to that of First View except for the following differences. The upper crystalline micrite unit of the Redwall is well developed. No karst topography is present. Chert-littered slopes at the base of the Watahomigi suggest the presence of basal conglomerate. At least six limestone ledges are present in the Watahomigi; no fossils were found in any of them. Some broad trough cross-stratification is present in the Manakacha cliff and several well-exposed bioturbated sandstones are present in the Watahomigi and Manakacha Formations. Local lateral variation can be observed in several units exposed in the area.

ROAD LOG—SECOND DAY: FLAGSTAFF TO GRAND CANYON, ARIZONA

Richard R. Rawson[1]

The field trip for the second day begins in Flagstaff, heads north around the east side of the San Francisco Volcanic Field, passes through Sunset Crater and Wupatki National Monuments, heads west at Cameron, Arizona, along the north side of the Gray Mountain Monocline to Desert View (eastern entrance to the Grand Canyon) and then along the South Rim of the Grand Canyon to the Visitors Center. The trip will afford views and discussions of Tertiary–Quaternary volcanic features, exposures of Paleozoic stratigraphy on the east side of the Mt. Elden Laccolith, monoclinally folded Paleozoic and Mesozoic strata, and views of Precambrian and Paleozoic stratigraphy in the Grand Canyon.

The total distance is approximately 125 miles. Arizona highway mileposts are noted where available outside of Grand Canyon National Park. The milepost markers are placed on each side of the highway facing traffic.

The road log begins on Switzer Mesa in Flagstaff—location of the United States Geological Survey Astrogeology Branch.

Total Mileage

0.0 **STOP 1—SWITZER MESA.** Discussion of Flagstaff's volcanic features. Proceed north to Cedar Avenue; turn right (east) and descend from Switzer Mesa which is capped by a basalt flow dated 5.8 m.y.

0.7

0.7 Road bends right and then left as it passes Coconino High School (on right). Elden Mountain at 9 o'clock. Note steep flow angles on dacite flows.

0.5

1.2 Stoplight. Turn right (south) onto 4th Street.

0.9

2.1 Stoplight. Turn left (east) onto U.S. 66. Permian Kaibab Limestone crops out across railroad tracks directly to south. Begin Arizona milepost system with 418.

0.7

2.8 Junction of U.S. 89 and 66. Keep to left on U.S. 89. Mt. Elden at 9 o'clock. Note longitudinal jointing in upper flow. Flow folds may be observed at the base of present Elden Mountain. Numerous volcanic cinder cones and cinder quarries on right. The cinders are used for cinder blocks, road material, and winter street maintenance. MP 418.7.

1.6

4.4 Camp Townsend. MP 420.3.

0.4

4.8 Leupp Road to right. Keep straight ahead. MP 420.7.

0.9

5.7 **STOP 2—MOUNT ELDEN VOLCANIC LACCOLITH.** MP 421.6.

0.7

6.4 Sunset Crater at 12 o'clock. MP 422.3.

0.4

6.8 Road to right leads to city dump and to a cinder field that was cratered by the U.S.G.S. to simulate a portion of the moon's surface for training astronauts and testing lunar equipment. Weights hanging from telephone cable on right side of road are to prevent standing waves during high winds moving through Schultz Pass (9 o'clock) which would break the cable. MP 422.7. Stay on U.S. 89.

4.6

11.4 Road ascends lava flow. Basalt in succeeding road cuts. MP 427.3.

1.8

13.2 View to left exposes inner basin thought to be a collapsed caldera of San Francisco Composite Cone. MP 429.1.

1.4

14.6 Entrance to Sunset Crater. MP 430.5 Turn right.

1.1

15.7 Road skirts edge of Bonito Park which was formed by damming of easterly drainage from San Francisco Peaks by Bonito lava flow that erupted from base of Sunset Crater.

0.9

16.6 Sunset Crater Visitors Center on right.

1.4

18.0 Turnout to left for photo stop of Bonito lava flow. The Bonito lava flow is somewhat unique, combining features of pahoehoe flows (lava tubes and hornitos) with typical rubbly aa surfaces. Collapsed lava tubes may be observed at this stop.

0.6

18.6 **STOP 3—SUNSET CRATER.**

0.8

19.4 Road now transversing an area covered with black cinders from Sunset Crater eruption.

4.7

24.1 Turn Left to **STOP 4—PAINTED DESERT VISTA.**

1.5

25.6 Numerous squeeze-ups dot lava flows around us as we travel north down into Little Colorado River valley. At 9 o'clock is Strawberry Crater, one of some 300 cinder cones in this field. Strawberry is heavily armored with spatter on its southwestern side.

1.3

---

[1] Department of Geology, Northern Arizona University, Flagstaff, Arizona 86011.

26.9 Continuing descent. At 3 o'clock is Haywire Crater, which has erupted several times. At 3:30 is rounded top of Merriam Crater, source of lava that formed Grand Falls, and launching site for hang gliders.

5.1

32.0 At 12:30 on near skyline is Woodhouse Mesa, capped by a basalt flow about 2.3 m.y. old. At 1:30 on far skyline is Black Point lava flow. Age of Black Point flow, which overlies a monoclinal structure by the same name, is 2.4 m.y.

1.8

33.8 Entering Wupatki National Monument. Brick-red sandstones and siltstones are Moenkopi Formation (Triassic). The monument preserves remnants of ancient pueblo structures constructed here by people of the Pueblo III culture. The earliest recorded human occupation of this area was about A.D. 600 by members of the Sinagua culture. After the eruption of Sunset Crater in A.D. 1065, Indians again returned to dry farm the area with its newly added black volcanic "mulch." The sites that remain in this area date from this later time, and illustrate an interesting relation between volcanic and human activity.

2.3

36.1 Wupatki Visitors Center. Rest stop. Continue north on Monument Road.

1.0

37.1 Rock unit underlying brick-red siltstones of Triassic Moenkopi Formation in this area is cherty to sandy, buff Kaibab Limestone of Permian age.

0.5

37.6 At 10 o'clock, Doney Crater. This crater is on a lineament that trends south and passes under Sunset Crater and O'Leary Peak. To the north the lineament extends into the Black Point monocline, which occupies the skyline at 3 o'clock. The Black Point monocline faces east, and Kaibab Limestone is exposed on the upthrown (western) side, with the Moenkopi Formation exposed on the downthrown (eastern) side.

1.5

39.1 Black Point monocline is partly a fold and partly a fault. Notice reddish Moenkopi remnants on fault plane, as well as patches of cinders, probably from eruption of Doney Crater.

0.5

39.6 To right, Kaibab Limestone. To left, interbedded cinders and bombs, along with fragments of underlying Kaibab Limestone and Coconino Sandstone lithologies.

0.4

40.0 San Francisco Peaks at 12:30. O'Leary Peak at 12 o'clock.

0.6

40.6 Turn left into View Point. In distance to east is the Little Colorado River valley, the Painted Desert country, and the Hopi Buttes diatremes. Return to road and turn left.

2.8

43.4 At 1 o'clock is SP Crater, a symmetrical cinder cone, with a rim armored by spatter. A lava flow issued from the base of the cone and flowed north, filling grabens associated with the Coconino Point–Gray Mountain monocline (on distant skyline at 3 o'clock). SP Crater has recently been dated at 70,000 years by the K-Ar method. The lava flow from SP Crater in aerial view is a familiar photograph in many text books.

3.9

47.3 Citadel Ruins to left.

0.4

47.7 Lomokai Ruins to right. Notice small fault in this area that offsets the Citadel basalt flow, as well as the Kaibab Limestone.

1.4

49.1 Crossing an area of numerous small grabens that interrupt the stripped Kaibab surface. To the west this stripped surface becomes a pediment developed on the Moenkopi along the monoclinal uplifts.

1.4

50.5 Highway on Black Point Basalt.

0.9

51.4 Road intersects U.S. 89; turn right (begin Arizona mileposts). Now driving north on a surface formed in Moenkopi Formation. Gray Mountain, on skyline from 6:30 to 10:30, is formed by the intersection of four monoclines that are a part of the East Kaibab Monoclinal system. Kaibab Limestone forms upper surface of monocline. Moenkopi rocks along highway and Kaibab on skyline both dip gently southward. MP 444.8.

1.2

52.6 Hanks Trading Post on left. MP 446.0.

4.6

57.2 Redbeds of Moenkopi Formation beneath basalt flows. MP 450.6.

2.1

59.3 Indian Trading Post. MP 452.7.

4.3

63.6 Village of Gray Mountain. MP 457.0.

1.0

64.6 **STOP 5—BLACK POINT MONOCLINE.** MP 458.0.

0.5

65.1 Driving north down stripped surface of Kaibab Limestone. At 3 o'clock is Black Point basalt flow. Ahead, near skyline, basal member of Triassic Chinle Formation, the Shinarump Conglomerate, is exposed in lower fold of Black Point Monocline. MP 458.5.

0.8

65.9 Contact with Kaibab and Moenkopi Formations. MP 459.3.

1.8

67.7 Contact of Shinarump and Moenkopi in roadcut. The Shinarump was formed by a braided stream system. Uranium has been found in this member throughout the Colorado Plateau associated with fossil logs which are locally abundant. MP 461.1.

2.3

70.0 Roadcut in Shinarump Conglomerate. MP 463.4.

1.5

71.5 Buck Rodgers Trading Post on right. MP 464.9.

0.3

71.8 Turn left (west) on Arizona 64 to Grand Canyon. Begin new milepost system with MP 296.0.

0.3

72.1 Roadcuts in upper part of Shinarump. At 2:30 is Shadow Mountain, the northernmost volcano in San Francisco Volcanic Field. It consists of interlayered tephra and flow units of alkaline basalt. MP 295.7.

0.9

73.0 Bridge crosses Tappan Wash. Tappan Wash basalt flow (0.5 m.y.) originated 60 miles (97 km) to south from an unknown source. Flow continues northward into the Little Colorado and then westward for about 9 miles (15 km). MP 294.8.

4.6

77.6 **STOP 6—COCONINO POINT MONOCLINE.** MP 290.2.

3.9

81.5 Photo stop. Turn right to parking lot. Overlook of Little Colorado River Gorge. Canyon about 1000 feet (305 m) deep at overlook. Kaibab Limestone overlies Coconino Sandstone. Toroweap Formation has changed facies here into a Coconino-like facies and cannot be distinguished from it. Several grabens may be observed. MP 286.3.

1.3

82.8 Side canyons cutting monocline expose near-vertical beds, in descending order, of Kaibab Limestone, Coconino Sandstone, and Schnebly Hill formation. MP 285.0.

3.2

86.0 Saddle Horn rock on left. The Shinarump makes the "Saddle Horn" and the balance of the exposures are in the Moenkopi Formation. Beautiful sedimentary structures are found in the Moenkopi at this site: ripple marks, mudcracks, burrows, and tracks. MP 281.8.

1.4

87.4 Viewpoint. Entrenched Little Colorado River has cut into marble platform.

1.9

89.3 Steeply dipping slickensided Kaibab Limestone exposed in roadcuts.

1.9

91.2 Road now on upper part of East Kaibab Monocline. At 3 o'clock is East Kaibab Monoclinal axis. At 10 o'clock Coconino Point Monocline continues west and becomes the Grandview Monocline.

0.5

91.7 Entering Kaibab National Forest.

8.6

100.3 Entering Grand Canyon National Park.

1.8

102.1 San Francisco Peaks visible to south.

1.1

103.2 Park Service checking station.

0.1

103.3 **STOP 7—DESERT VIEW.** Return to main road and proceed west.

0.8

104.1 Turn right. **STOP 8—LIPAN POINT.** Return to main road, turn right.

1.3

105.4 Side road to left leads to Tusayan Ruins and Museum. Pueblo-type Indians built the Tusayan structures about A.D. 1185.

3.8

109.2 Side road to right leads to Moran Point.

3.0

112.2 Highway now ascending Grand View Monocline with a relief of 400 feet (120 m) on Kaibab surface.

3.0

115.2 Turn right to **STOP 9—GRAND VIEW POINT.** Return to main road, turn right.

7.6

122.8 Side road to right leads to Yaki Point and head of Kaibab Trail which you may descend tomorrow to walk through Redwall Limestone and Supai Group section.

1.1

123.9 Junction of Rim Drive with south entrance road. Turn right.

0.7

124.6 Mather Point on right.

0.6

125.2 Road to right leads to Yavapai Museum and Yavapai Point.

0.8

126.0 Visitors Center, Grand Canyon National Park.

END OF SECOND DAY ROAD LOG

STOP DESCRIPTIONS—SECOND DAY

Richard R. Rawson[1]

STOP 1—SWITZER MESA—UNITED STATES GEOLOGICAL SURVEY FIELD CENTER
Sec. 10, T. 21 N., R. 7 E.
Coconino County, Arizona

Turn and face north (toward Peaks). The San Francisco Peaks form the skyline directly to the north. The highest peak, Mt. Humphrey, is 12,680 feet (3,855 m) and hidden by Mt. Agassiz (far left) from this viewpoint. The Peaks are a composite volcano composed of rhyolite, rhyodacite, dacite, andesite flows, and breccias. The lower and closer hills below the Peaks at 12 o'clock are the Dry Lake Hills composed of dacite. Mt. Elden is located at 1 o'clock and composed of high-angle dacite flows. To the east of Mt. Elden a few low-lying volcanic cinder cones may be observed. We are standing on a basalt flow that originated to the north and flowed southward, capping this mesa that predates the San Francisco Peaks and Mt. Elden. Other similar flows may be observed to the south at Anderson Mesa and to the west. Lowell Observatory, where Pluto was discovered, is located on the mesa farthest to the west. The basalt flows cap the Triassic Moenkopi Formation. Post-basalt erosion in areas not covered by the basalt has cut down to a broad stripped surface on the Permian Kaibab Limestone. The oldest rocks in the San Francisco Peaks have been dated as 2.7 m.y. and the youngest (Sugarloaf) have been dated at 0.2 m.y.; Dry Lake Hills, 0.8 m.y.; Mount Elden, 0.5 m.y.; Anderson Mesa, 6.0 m.y.; Switzer Mesa, 5.8 m.y.

STOP 2—MT. ELDEN LACCOLITH (?)
Sec. 28, T. 22 N., R. 8 E.
Coconino County, Arizona

At 3 o'clock three conical knobs composed of Permian Kaibab Limestone dip 50° to the east. Other Paleozoic rocks underlying the Kaibab in descending order toward Mount Elden are Permian Coconino Sandstone, Permian Schnebly Hill formation (Blakey, this guide book). Permian Hermit Shale, Pennsylvanian-Permian Supai Group, Mississippian Redwall Limestone, and the Devonian Martin Formation which rests on dacite high up on Mt. Elden. Another block of Paleozoic strata is exposed on the western side of Mt. Elden dipping 16° to the northwest (fig. 3). The blocks were uplifted during the formation of Mt. Elden. Robinson (1913) described this as a laccolith. Sunset Crater (Stop 3) is at 11 o'clock, O'Leary Peak at 10:30, and Sugarloaf Mountain (rhyolite dome) at 10 o'clock. O'Leary Peak has a large block of Coconino Sandstone at its top rafted up by the rhyodacite that forms most of the domal uplift.

STOP 3—SUNSET CRATER
Sec. 23, T. 23 N., R. 8 E.
Coconino County, Arizona

Sunset Crater was named by J. W. Powell (Colorado River explorer) because the color of the red cinders surrounding the top of the crater appeared as if caused by the setting sun. The volcano is the youngest in the San Francisco volcanic field, having been dated as A.D. 1065 by tree-ring chronology. The cinder cone rises 800 to 1,000 feet (250–300 m) above the surrounding basalt flows and cinder deposits. Two basaltic flows were erupted from the base of this volcano—the Bonito flow to the west and the Kana-a flow to the northeast. Several small spatter cones are visible near this stop. Notice that some of these are aligned along a fissure that heads toward Sunset Crater. Xenoliths of Paleozoic rocks occur in the basalt. The high peak toward the north is O'Leary Peak, composed of rhyodacite dated as 0.23 m.y., making this the youngest silicic eruption in the San Francisco Field. Sunset Crater became a National Monument as the result of local concerned citizens protesting movie companies "staging" eruptions by setting off dynamite and burning old tires in the cone.

STOP 4—PAINTED DESERT VISTA
Sec. 10, T. 23 N., R. 9 E.
Coconino County, Arizona

The color in the Painted Desert is provided by the Triassic Moenkopi and Chinle Formations and the Jurassic Moenave and Kayenta Formations to the northeast. To the north and east is the valley of the Little Colorado River. The Hopi Buttes may be observed (on a clear day) 80 miles (130 km) to the northeast. The Buttes are volcanic necks formed of kimberlite (sorry, no diamonds). The cinder cones and lava flows in the immediate area yield age dates as old as 3 m.y. Most of the flows moved down drainage toward the Little Colorado River and one from Merriam Crater reached the river and dammed it, causing the river to run around the flow and then fall back into its old channel forming the Grand Falls.

STOP 5—BLACK POINT MONOCLINE
Sec. 6, T. 27 N., R. 9 E.
Coconino County, Arizona

At this fold the prevailing gentle south dip has been reversed to a moderate north dip on the Black Point Monocline. The fold continues eastward 8 to 10 miles (12–15 km) and then makes a right angle turn to the south at the Black Point flow and passes beneath Doney Crater. To the west the monocline intersects the Coconino Point and additional hill monoclines in a zone of underthrusting and tear faulting. The Kaibab on Gray Mountain is 2,000 feet (610 m) higher than the Kaibab at the base of the monocline. Looking northward down the dip of the monocline, the red Moenkopi Formation may be observed near the base of the fold. The light-colored unit capping the Moenkopi Formation is the Shinarump Conglomerate, the basal member of the Triassic Chinle Formation. At 10:30 is Shadow Mountain volcano. Dates on flows from the volcano indicate eruption about 0.5 m.y. ago.

---

[1] Department of Geology, Northern Arizona University, Flagstaff, Arizona 86011.

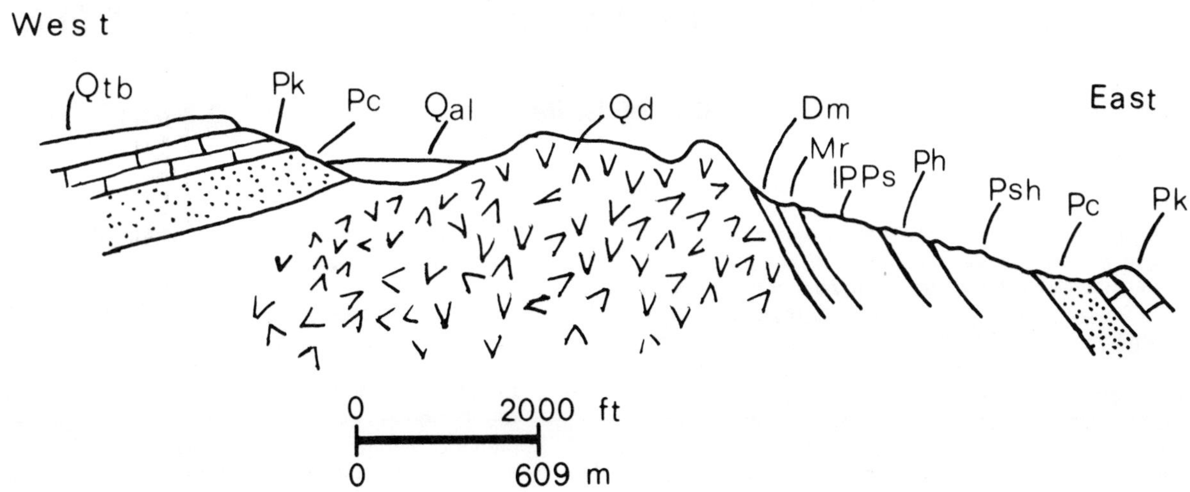

Figure 3. Cross-section, Elden Mountain. No vertical exaggeration. Qal – alluvium; Qd – dacite; Qtb – basalt flow; Pk – Kaibab Limestone; Pc – Coconino Sandstone; Psh – Schnebly Hill formation; Ph – Hermit Shale; PPs – Supai Group; Mr – Redwall Limestone; Dm – Martin Formation. (After Barnes, Eastwood, and Beus, 1974)

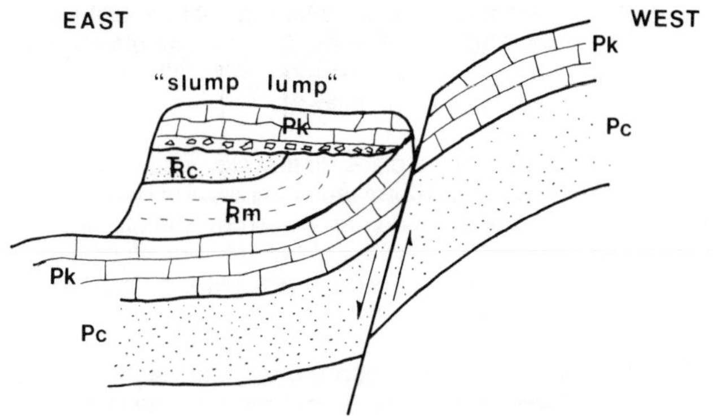

Figure 4. Cross-section across "slump lump" at Pothole Canyon. Trc – Chinle Formation; Trm – Moenkopi Formation; Pk – Kaibab Limestone; Pc – Coconino Sandstone. (After Barnes, Eastwood, and Beus, 1974)

STOP 6—COCONINO POINT MONOCLINE
Sec. 22, T. 28 N., R. 8 E.
Coconino County, Arizona

The westward turning of the Coconino Point Monocline may be seen from this stop. The steep dips of Kaibab and resulting flatirons become prominent along some portions of the fold. Overturning of the Kaibab has been documented in some sectors. At 10 o'clock a feature locally known as "slump lump" (fig. 4) may be seen. A large flatiron of Permian Kaibab broke loose and glided down the monocline to cover the upturned edges of the Triassic Moenkopi and Chinle Formations near the lower fold axis, and becomes apparently conformable away from the axis.

STOP 7—DESERT VIEW
Sec. 23, T. 29 N., R. 8 E.
Coconino County, Arizona

The Colorado River may be seen in the lowest part of the Canyon where it is cutting the younger Precambrian rocks. The Dox, Cardenas, and Nankoweap Formations are beveled by the "Great Unconformity" and overlain by the Cambrian Tapeats Sandstone (see fig. 5). Rocks of Cambrian, Devonian, Mississipian, Pennsylvanian, and Permian age form the Paleozoic section. The Canyon is over 4,500 feet (1,380 m) deep at this location. The Kaibab Limestone forms the rim rock in all directions. The East Kaibab Monocline may be viewed to the east. Three miles (5 km) to the east an erosional remnant (Cedar Mountain) is formed by the Moenkopi Formation and the Shinarump Conglomerate.

STOP 8—LIPAN POINT
Sec. 28, T. 31 N., R. 5 E.
Coconino County, Arizona

From this viewpoint the entire Grand Canyon section can be seen from the oldest rocks, Vishnu Schist (1.8 b.y.), to the Permian Kaibab Limestone.

STOP 9—GRAND VIEW POINT
Sec. 8, T. 30 N., R. 4 E.
Coconino County, Arizona

About 2,500 feet (765 m) below the viewpoint is Horseshoe Mesa, formed by the Mississippian Redwall Limestone. This is the type section for the Horseshoe Mesa Member of the Redwall. In the 1890's, the Last Chance Copper Mine was producing from the Redwall Limestone on Horseshoe Mesa. Ore was hauled to the rim by mules.

TRAIL LOG—THIRD DAY: SOUTH KAIBAB TRAIL, GRAND CANYON, ARIZONA

Stanley S. Beus[1]

This trip will begin at the Kaibab Trail head at Yaki Point, 4 miles (6 km) east of the Visitors Center on the South Rim, Grand Canyon National Park. We will traverse the upper two-thirds of the Paleozoic stratigraphic sequence beginning in the Permian Kaibab Limestone that forms the canyon rim and ending at the base of the Mississippian Redwall Limestone which forms the prominent vertical cliffs at about mid-point in the Grand Canyon walls. There will be opportunity to examine the four newly named members of the Supai Group (McKee, 1975) and the four members of the Redwall Limestone, in addition to the Permian strata at the canyon rim (fig. 5).

The trail distance to take us through the Redwall Limestone is about 3.8 miles (6 km) and involves a descent from 7,000 feet (2,100 m) to about 3,400 feet (1,035 m), a vertical distance of 3,600 feet (1,100 m). The return journey up the trail can be very exhausting, especially on a hot day. You may wish to proceed only part of the way down the trail. Water will be available at the Cedar Ridge rest stop (top of the Supai Group) for the return trip. All should carry at least a 2-quart (2-liter) supply of water.

Miles:

0.0 Begin descent of Kaibab Trail through limestone ledges of Kaibab Formation. Kaibab Limestone forms a steep 60–m (200–ft) cliff at canyon rim.

Third Switchback corner: note brown ellipsoid or irregular chert nodules parallel to bedding in Kaibab.

Fifth switchback corner: fossil productid brachiopods and sponges abundant in cherty limestone beds of the Kaibab.

Sign: Kaibab-Toroweap contact. The Toroweap has three distinct members: the upper member, 53 m (175 ft) thick, forms a slope and is composed of sandstone and siltstone, including some red beds and thin limestones; the middle member is a resistant ledge-forming limestone 21 m (70 ft) thick; and the lower member is a cross-bedded to flat-bedded quartz sandstone 33 m (110 ft) thick.

Hugh slump blocks of Kaibab limestone lie scattered on the Toroweap slope. Specimens of brachiopods and sponges locally occur on weathered surfaces of the slump blocks.

0.05 Cross-bedded quartz sandstone beside trail is in the upper member of the Toroweap. Fifteen-meter (50–ft) vertical cliff of gray-brown limestone is the middle member of the Toroweap.

Sign: Toroweap-Coconino contact; clearly marked by color boundary between red-brown flat-bedded sandstone of the Toroweap and cross-bedded sandstone of the Coconino, which forms a massive 100–m (350–ft) buff-yellow cliff of canyon wall.

Viewpoint just beyond corner of 10th switchback affords panoramic view of trail ahead and entire Paleozoic section down to Cambrian beds along the Tonto Platform at top of Inner Gorge. Large-scale cross-beds up to 9 m (30 ft) high in Coconino Sandstone are mainly tabular-planar or wedge-planar shape with predominant southerly dips. A few bedding planes display gentle broad ripple marks, and in other localities of the Grand Canyon, vertebrate trackways.

1.0 Sign: Coconino-Hermit contact; actual contact largely obscured. The Hermit Shale forms a smooth slope on deep-red siltstone and shale beneath cliffs of Coconino and is about 90 m (300 ft) thick.

Cedar Ridge. Good place for rest stop where trail enters juniper-covered platform area near base of Hermit Shale. View of plant fossils under glass-covered display case about 65 m (75 yd) to left of trail. Mud-crack casts, rain prints, fossil ferns, conifers, and insects occur in this part of the Hermit.

Hermit-Supai contact about where trail begins descent from Cedar Ridge. The Supai Group is more than 300 m (1,000 ft) of dark-red to orange-brown sandstone, siltstone, and limestone of Pennsylvanian and Permian age. The system boundary is approximately one-quarter the distance down from the top of the Supai in the slope beneath the massive upper cliff of Esplanade Sandstone (figs. 6, 7). Irregular limestone pebble conglomerate marks Permian-Pennsylvanian boundary and contact between Esplanade and Wescogame.

Trail on saddle between Cedar Ridge and O'Neill Butte in Wescogame Formation.

1.5 Trail descends gradually through the Supai, passes by east side of O'Neill Butte.

2.0 Pennsylvanian reptile or amphibian trackway in sandstone beds to left side and about 1 m (3 or 4 ft) above trail, lower Wescogame of the Supai Group. Wescogame-Manakacha boundary is near the base of the cliff in which these tracks occur.

Trail on ridge crest.

2.5 Trail in Watahamogi Formation crosses some blue-gray finely crystaline limestone. Pennsylvanian fusulinids and brachiopods occur in similar limestones as far east as Marble Canyon (E. D. McKee, 1978, written commun.,). East of the trail can be seen the onlapping Tapeats against a promontory of Zoroaster Granite (Precambrian), in the Inner Gorge across the Colorado River.

[1]Department of Geology, Northern Arizona University, Flagstaff, Arizona 86011.

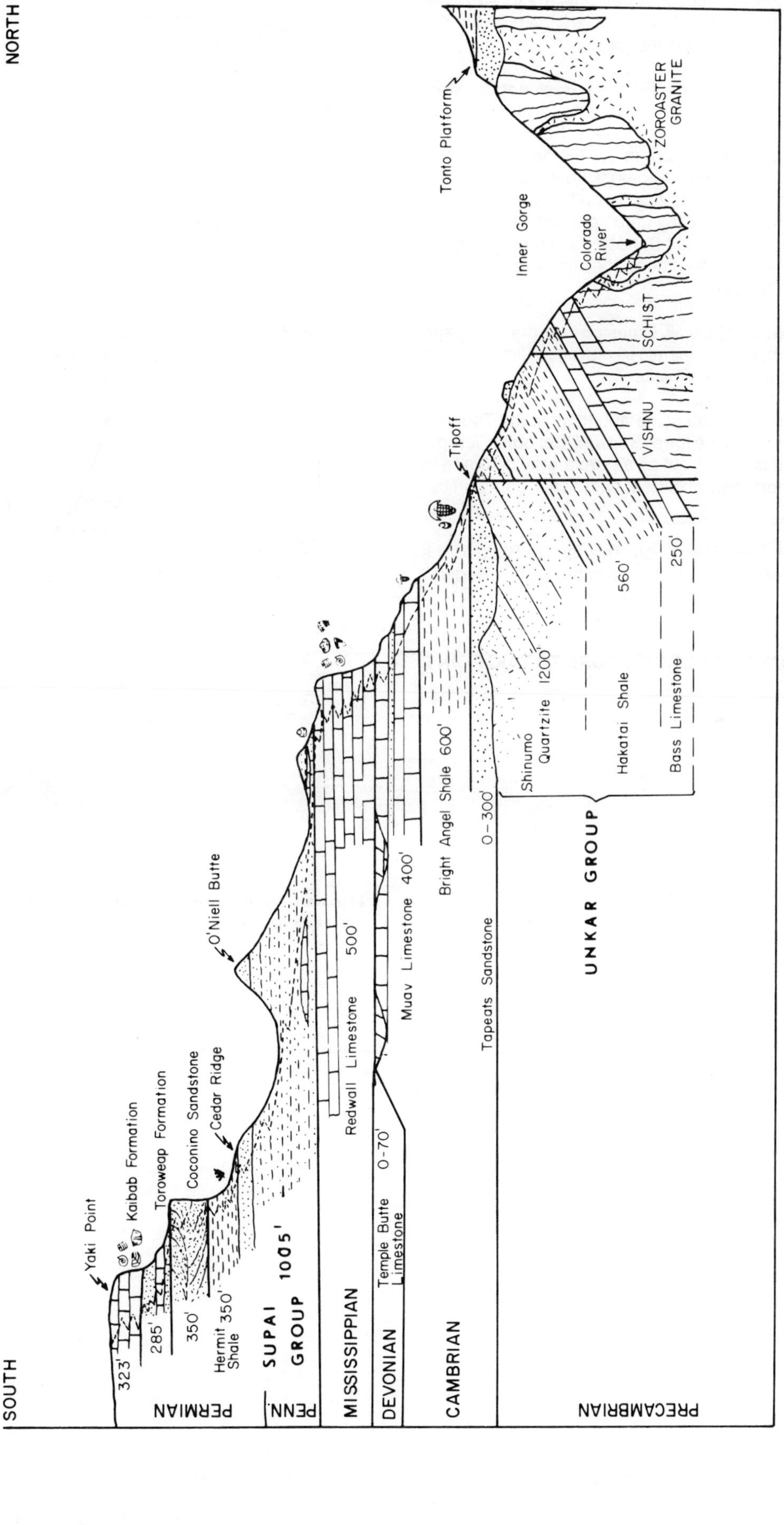

Figure 5. Cross-section of south wall of Grand Canyon approximately at position of South Kaibab Trail.

3.0 Trail begins descent through sheer 150–m (500–ft) cliffs of Redwall Limestone of Mississippian age (see fig. 66, Rawson and Kent, this guidebook).

Third switchback corner in Redwall. Rest stop for view across Canyon to "Cheops Bay" on Tonto Platform across the river. Cambrian Tapeats Sandstone beds (red-brown rounded ledges), in the flat east and below Cheops Pyramid, pinch out against promontories of Precambrian Shinumo Quartzite on both sides of a mile-wide Tapeats outcrop. Gray-green patches of Bright Angel Shale rest on Tapeats Sandstone in the center, but on Shinumo Quartizite around the margins of this ancient Cambrian "bay."

Solution cave in Redwall Limestone to the right and above the trail; white minerals filling vugs in limestone at trail level are calcite and barite.

3.5 Chert and crinoidal limestone abundant in the Thunder Springs Member of the lower Redwall. Devonian Temple Butte Formation crops out as a purple-gray dolomite in the cliff just northeast of the last switchback corner in the Redwall.

Sign: Temple Butte Formation. The Temple Butte forms thin lenses of channel deposits between the Redwall and the Cambrian Muav Limestone.

END OF THIRD DAY TRAIL LOG; RETURN TO YAKI POINT

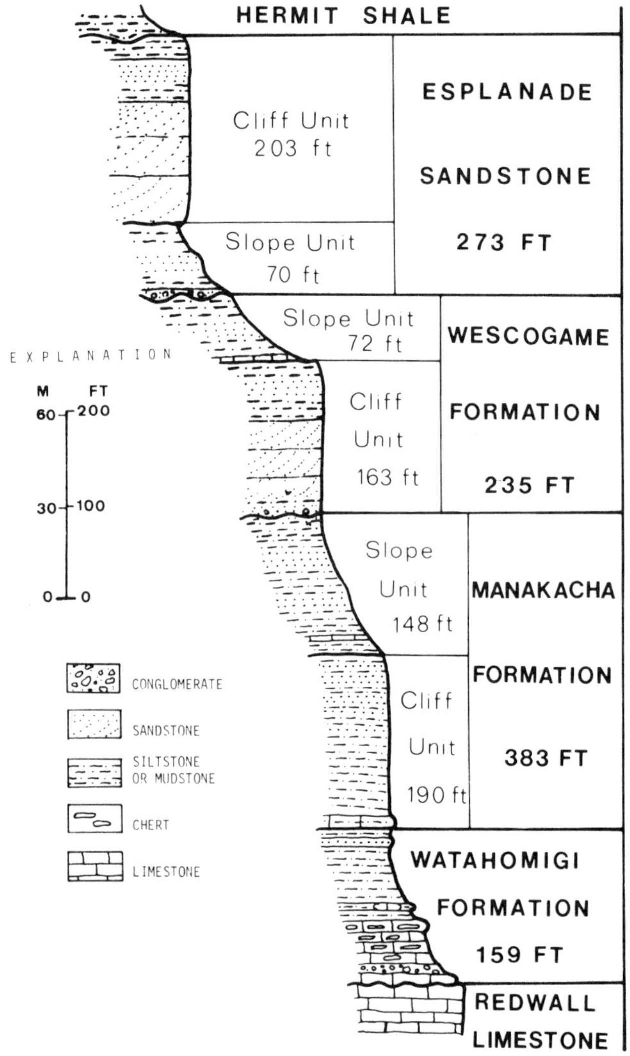

Figure 6. Columnar section of Supai Group along South Kaibab Trail. Thickness data provided by E.D. McKee (1978, written commun.).

Figure 7. View of Supai Group along South Kaibab Trail beneath O'Neill Butte. Photograph taken at about milepost 0.7. R – Redwall Limestone; W – Watahomigi Formation; M – Manakacha Formation; We – Wescogame Formation; E – Esplanade Sandstone. Dashed lines mark approximate formation boundaries.

ROAD LOG — FOURTH DAY: GRAND CANYON TO KINGMAN, ARIZONA

Stanley S. Beus,[1] Richard R. Rawson,[1] and Ronald C. Blakey[1]

This excursion begins at the south rim of Grand Canyon, passes through Williams, Ash Fork, and Seligman, and ends at Kingman, Arizona. The trip will afford visits to the Pennsylvanian-Permian boundary in the Supai Group at Chino Point, west of Seligman, a cave in the Redwall Limestone, the Devonian-Mississipian boundary at Peach Springs, and distant views of Paleozoic stratigraphy at the southern margin of the Colorado Plateau (fig. 8).

The total distance is approximately 190 miles. Arizona highway mileposts are noted where available outside of Grand Canyon National Park.

Total Mileage

0.0 Visitors Center in Grand Canyon National Park. Proceed northeast on West Rim Drive.

0.4

0.4 Junction with Yavapai Point road on left. Continue on West Rim Drive.

0.6

1.0 Mather Point on left.

0.8

1.8 Junction with East Rim Drive (Arizona 64) on left. Proceed south on 64.

1.4

3.2 South Rim entrance station, Grand Canyon National Park. Road is on Kaibab Limestone surface for next 35 miles (56 km).

3.2

6.4 South boundary of national park.

0.2

6.6 Moqui Lodge on right.

1.0

7.6 Tusayan; major tourist accommodations and staging area for helicopter trips over and through Grand Canyon. Begin Arizona milepost 236.

1.4

9.0 Grand Canyon Airport road to right. MP 234.6.

5.3

14.3 Roadcut in Kaibab Limestone. MP 229.3.

1.3

15.6 Road on right leads to Anita Station, site of an early copper mine to which a mining railroad spur from the Sante Fe was built in 1898. The copper ore played out quickly; the Sante Fe bought the new mining railroad and extended it to Grand Canyon in 1901, initiating extensive tourist travel to the Grand Canyon by rail. MP 228.0.

3.9

19.5 Leaving Kaibab National Forest. Prominent landmark to left (east) is Red Butte, an outlier of Triassic redbeds (Moenkopi and Shinarump) capped by Tertiary basalt dated at 8.0 m.y. MP 224.1.

2.6

22.1 Road on right to Willaha, site of an abandoned copper mine. MP 221.5.

1.5

23.6 Panoramic view of major composite volcanoes of the San Francisco Peaks 12,655 feet (3,400 m), Kendrick 10,418 feet (3,100 m), Sitgreaves 9,388 feet (2,854 m) and Bill Williams Mountain 9,255 feet (2,814 m). MP 220.0.

6.4

30.0 Junction with U.S. 180 to Flagstaff on left (keep right). Bedrock City on right. Bill Williams Mountain at 12 o'clock. MP 213.6.

6.9

36.9 Roadcut in basalt, edge of San Francisco Volcanic Field. MP 206.7.

0.4

37.3 Elongage cinder cone and dike(?) on left. MP 206.3.

0.6

37.9 Roadcuts in Tertiary volcanics for next 25 miles. MP 205.7.

0.9

38.8 Pump station for pipe line on left. MP 204.8.

3.8

42.6 Tammyville Campground on left. MP 201.0.

2.0

44.6 Cross power line. MP 199.0.

1.0

45.6 Cinder cones on left. MP 198.0.

4.0

49.6 Cedar Mountain on left 3 miles to east; Kendrick, San Francisco Peaks and Sitgreaves in distance. MP 194.0.

0.5

50.1 Abandoned Grand Canyon spur of Santa Fe Railroad on right. MP 193.5.

2.4

52.5 Spring Valley Road on left, stratified cinders and ash in cinder pit to left. MP 191.1.

0.9

53.4 Canyon Trading Post on right. MP 190.2.

3.9

[1] Department of Geology, Northern Arizona University, Flagstaff, Arizona 86011.

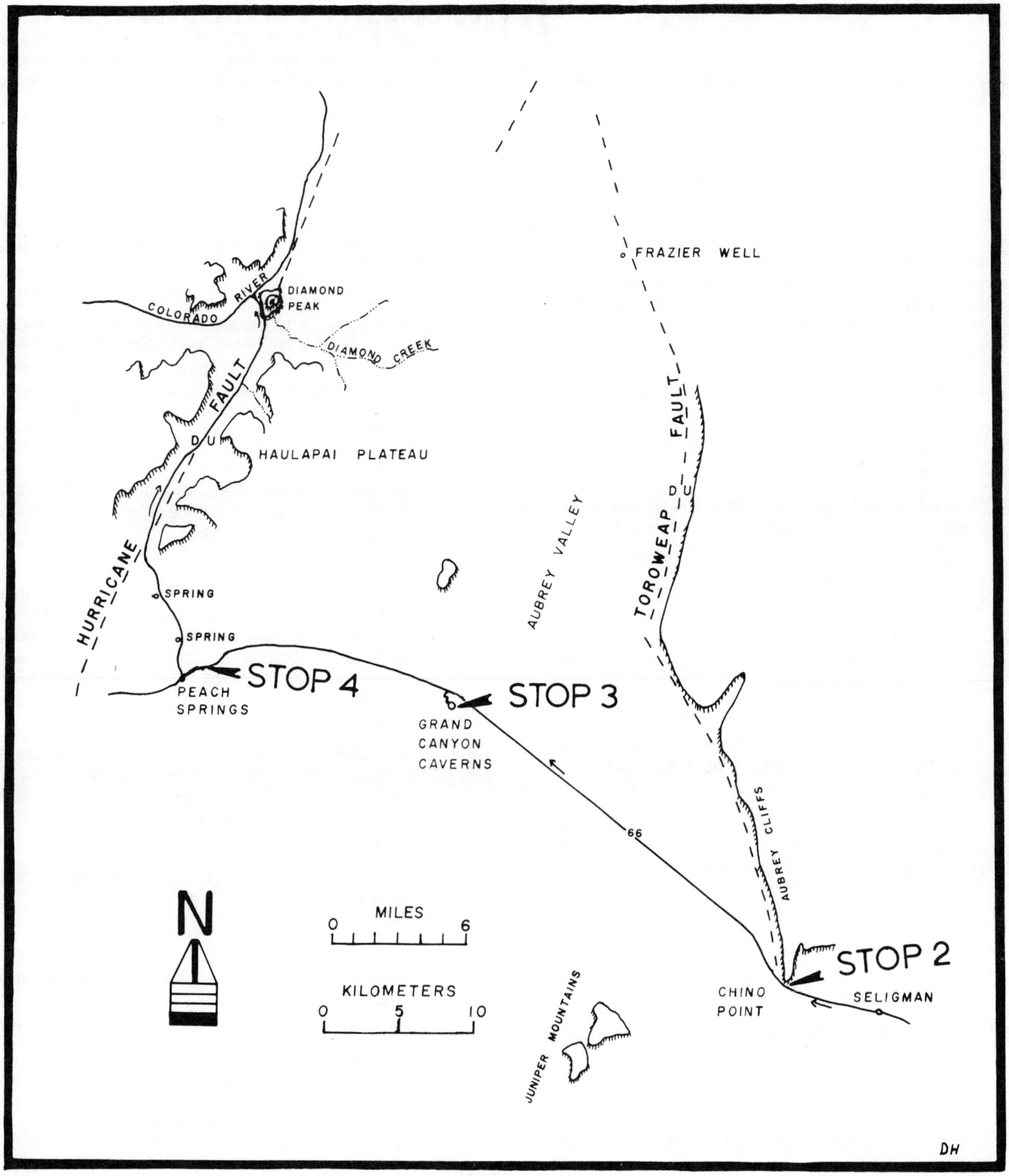

Figure 8. Index map showing locations of Stops 2-4, fourth day road log.

57.3 Road to Kaibab Lake on right. MP 186.3.

1.0

58.3 Junction with U.S. 66 (I-40). Turn right (west) toward Williams on 66. Begin Arizona mileposts on 66 with MP 166.0.

0.6

58.9 Underpass beneath Santa Fe Railroad. MP 165.4.

0.5

59.4 Overpass over Santa Fe Railroad; enter Williams, Arizona. To southwest is Bill Williams Mountain, a late Cenozoic composite volcano named for Bill Williams (William Shirley Williams), a famous hunter, trapper, and mountain man who visited Arizona in 1837. MP 164.9.

1.9

61.3 Leave Williams. MP 163.0.

1.0

62.3 Roadcut on right in a cinder cone. MP 162.0.

2.0

64.3 Cinders, volcanic bombs, dikes, and basalt flows in roadcut on right. MP 160.0.

2.8

67.1 McLellan Reservoir on right skirted by abandoned railroad. MP 157.2

1.1

68.2 Red volcanic cinders and black basalt flows in roadcut on right. MP 156.1.

1.1

69.3 Road begins descent of Ash Fork Hill; basalt, tuff and agglomerate beds in roadcut on right. MP 155.0.

1.5

70.8 Juniper-covered mesas to right (north) capped with Cenozoic basalts overlying Permian limestone. Prominent buttes ahead (west) are Picacho Butte on the left (7,250 ft (2,204 m)) and Mt. Floyd on the right (7,441 ft (2,262 m)). Downgrade on highway is on basalt ramp over Permian Kaibab and Toroweap Formations and Coconino Sandstone. Mogollon Rim becomes indistinct in this area. MP 153.5.

2.5

73.3 Escarpment to left (southwest) is dip slope and outliers of late Paleozoic rocks. Closer smooth skyline is Mogollon slope with Redwall Limestone and outliers of Supai Group. More irregular distant skyline is Juniper Mountains composed of approximately lower two-thirds of Supai Group. Distant peaks to south are Precambrian granite. MP 151.0.

5.0

78.3 Entering Ash Fork, "Flagstone capital of the world." In the late 19th century Ash Fork was a major freight and passenger station connecting with the stage line to Phoenix, and boasted a large Fred Harvey restaurant. Escarpment to the right at 4 o'clock is Permian Toroweap and Kaibab Formations overlain by volcanics in places. The red-brown slope beneath is scarred with numerous quarries for flagstone from the Coconino Sandstone which underlies the Toroweap. MP 146.0.

0.7

79.0 On the right, note stock piles of flagstone from the Coconino Sandstone and other decorative stone at Kaibab Stone Company, Ash Fork. Stone from here is shipped all over the western states for fireplace and building facing. MP 145.3.

0.5

79.5 Roadcut in Cenozoic volcanics on right. MP 144.8.

0.5

80.0 Railroad overpass. Juniper-covered hills to left (southwest) are all in Supai Group (Pennsylvanian-Permian). MP 144.3.

1.3

81.3 Picacho Butte; at 12 o'clock on the skyline is an andesitic Tertiary dome. Mt. Floyd, the highest of the cluster of peaks at about 2 o'clock is likewise a Tertiary intrusive. Next several roadcuts are in late Cenozoic basalts which underly the floor of Partridge Creek Valley for several miles. MP 143.0.

0.5

81.8 Bridge over Partridge Creek. Abandoned railroad right-of-way on right is the old Santa Fe. The new track is several miles to the north. MP 142.5.

0.5

82.3 Escarpments to left (south) at 9 o'clock are mainly of Mississippian Redwall and Devonian Martin Formations. Reddish tree-covered buttes are Supai Group. Chino Valley, beyond the near escarpments, marks southwest edge of Colorado Plateau in this area. MP 142.0.

1.0

83.3 Juniper Mesa at 10 o'clock, the distant mesa across Chino Valley, is capped with Redwall Limestone. Picacho Butte at 1 o'clock on skyline. Mt. Floyd and related intrusive plugs at 2 o'clock on skyline. MP 141.0.

1.0

84.3 Crookton exit. Roadcut in late Cenozoic basalt. Transwestern and El Paso Natural Gas pipelines parallel roadway on right. MP 140.0.

2.1

86.4 Roadcut in late Cenozoic basalt. Columnar basalt on left-center of cut. MP 137.9.

0.9

87.3 Roadcut through dark-purplish ash. Rolling country to north is karst topography on Redwall Limestone with thin cover of basalt. Some sinkholes are reportedly several hundred feet deep. MP 137.0.

0.2

87.5 Roadcuts in cinders, volcanic ash, and basalt, next few miles. Picacho Butte to left front at 10 o'clock. MP 136.8.

1.8

89.3 Cluster of peaks at 2 o'clock (northwest) are lava domes in late Cenozoic Mt. Floyd Volcanic Field. MP 135.0.

3.0

92.3 Summit between Partridge Creek and Chino Valleys. Picacho Butte directly south at 9 o'clock has long been a landmark in northwestern Arizona. Tertiary intrusive rocks form the high point. On south and west flanks is a nearly complete Paleozoic section which includes Tapeats, Bright Angel, Martin, Redwall, Supai, Hermit, and Coconino Formations. This section is critical for correlation of Supai Group and related rocks into the Mogollon Rim. To right (north) is Mt. Floyd, another volcanic feature. Roadcuts are in basalt and cinders in next 5 miles (8 km). MP 132.0.

4.1

96.4 **STOP 1 — CHINO VALLEY OVERLOOK** and landmarks to the west. MP 127.9.

2.8

99.2 Roadcut in Tertiary basalt. MP 125.1.

1.4

100.6 Seligman exit 123, stay on I–40. MP 123.7.

2.3

102.9 Take exit 121 to right toward Seligman. End of I–40 mileposts.

0.4

103.3 Cross wash.

0.4

103.7 Railroad overpass.

0.5

104.2 Junction with U.S. 66 at west edge of Seligman. Turn left (west) on 66. Seligman is a declining tourist and railroad town that once served as the junction of the railroad from Prescott (now abandoned) with the main Sante Fe line. Begin U.S. 66 mileposts with MP 141.3.

0.2

104.4 Underpass beneath I–40 exit. MP 141.1.

1.7

106.1 Cross Chino Wash. Promontory at 12 o'clock, north side of road, is Chino Point. At road level is upper Supai Group beneath a thin section of Hermit Shale, Toroweap and Kaibab Formations above. MP 139.4.

3.0

109.1 **STOP 2 — CHINO POINT.** Pennsylvanian – Permian boundary in Supai Group. MP 136.4.

0.4

109.5 Lower red-brown ledges on right (north) represent top of Supai. Smooth slope above is Hermit Shale and Coconino Sandstone with Toroweap and Kaibab Formations on crest. Escarpment to left at 9 to 12 o'clock composed of middle Paleozoic carbonates, mainly Devonian Martin Formation and Mississippian Redwall Limestone. MP 136.0

1.0

110.5 Roadcut in Mississippian Redwall Limestone dipping northeast at 20-30°. Redwall and Devonian Martin Formation form escarpment to south. Permian section is to north on Aubrey cliffs. Monocline and/or fault separate the two sections. MP 135.0.

1.0

111.5 Entering Aubrey Valley. Escarpment on right of late Paleozoic formations forms the Aubrey Cliffs which are partially the result of high-angle faulting along southward extension of Toroweap fault. MP 134.0.

1.0

112.5 Sign: Grand Canyon Caverns 17 miles. Railroad cut to left at 9 o'clock in Redwall Limestone outcrops. MP 133.0.

2.0

114.5 Road traversing Quaternary sediments that floor Aubrey Valley. This part of Aubrey Valley is a closed basin. Escarpments on right (northeast) are Aubrey Cliffs capped with Permian Kaibab Limestone. Summits to left (southwest) at about 9 o'clock are northern end of Juniper Mountains capped by Mississippian Redwall Limestone and locally by basalt. Structural relief is on the order of 1,000 feet (305 m). MP 131.0.

8.0

122.5 Road junction with Pica Camp Road, on right. MP 123.0.

3.0

125.5 Road begins gradual ascent of western side of Aubrey Valley and passes from Quaternary sediments to Devonian and Mississippian limestones and dolomite. Roadcut here is in Redwall Limestone. MP 120.0.

3.0

128.5 Hyde Park Village. Note orange tank at left (southwest) over Grand Canyon Caverns. MP 117.0.

1.0

129.5 Roadcut in Redwall Limestone; note karst-like weathering. MP 116.0.

0.5

130.0 **STOP 3 — GRAND CANYON CAVERNS.** MP 115.5.

1.5

131.5 Quaternary volcanic features on right (north); low hills to left front at 11 o'clock are middle Paleozoic limestones (Devonian and Mississippian); roadcut in Redwall. MP 114.0.

0.5

132.0 Enter Hualapai Indian Reservation. MP 113.5.

1.5

133.5 Hills of Devonian Martin Formation to left (south). MP 112.0.

0.3

133.8 Roadcut in Quaternary alluvium. MP 135.8.

0.2

134.0 Roadcuts in Redwall. MP 111.5.

1.0

135.0 Junction with road to Havasu Canyon and Frazer Well on right. MP 110.5. Continue on U.S. 66.

0.4

135.4 Redwall Limestone in roadcut to left. MP 110.1.

0.7

136.1 Entering Mohave County. MP 109.4.

1.1

137.2 Roadcut on right in deeply weathered and brecciated Redwall Limestone and cemented gravels. MP 108.3.

0.8

138.0 Roadcut in Redwall Limestone overlain by partially cemented Quaternary alluvium. MP 107.5.

0.9

138.9 Roadcut in Redwall Limestone. MP 106.6.

1.0

139.9 Roadcuts in Redwall Limestone. MP 105.6.

0.1

140.0 Begin descent of grade into Peach Springs. Road is cut into Redwall Limestone (Mississippian), Martin Formation, mainly dolomites (Devonian), Cambrian undifferentiated dolomite, and Muav Limestone (Cambrian) in successive order in the next two miles. MP 105.5.

0.3

140.3 **STOP 4 — DEVONIAN–MISSISSIPPIAN CONTACT NEAR PEACH SPRINGS** in roadcut on left (south). MP 105.2.

0.2

140.5 View to right (north) of head of Peach Springs Canyon cut into lower Paleozoic strata. MP 105.0.

0.3

140.8 Roadcut in Cambrian undifferentiated dolomite and/or Upper Muav on left. MP 104.7.

0.7

141.5 Thin-bedded gray limestone of Muav Formation in roadcuts on left and right. MP 104.0.

0.5

142.0 Entering Peach Springs. Hills to south of highway expose carbonate strata of Cambrian Muav Limestone, Cambrian undifferentiated, and Devonian Martin Formation. MP 103.5.

0.5

142.5 Leaving Peach Springs. Roadcut in basalt on right. MP 103.0.

1.2

143.7 Cherokee Point to left (south). Section ranges from Cambrian Muav Limestone to Mississippian Redwall. Smooth slope is Cambrian undifferentiated. MP 101.8.

1.8

145.5 Entering Truxton Valley near edge of Colorado Plateau. Jagged peaks on skyline are Hualapai Mountains composed of Precambrian granite and metamorphics. Kingman lies at foot of these mountains. Although they may appear close, they are 50 miles away. MP 100.0.

5.0

150.5 Entering Truxton. Music Mountain to right (northwest) exposes Precambrian through Devonian rocks. Prominent limestone cliff is Rampart Cave Member of Muav Limestone; lower light-colored slope is Precambrian granite capped by Cambrian Tapeats Sandstone which forms bench. MP 95.0.

4.0

154.5 Roadcuts in Tertiary volcanics with tuffs, ash flows, cinders, and miscellaneous ejecta. MP 91.0.

2.0

156.5 Butte to right (north) is capped by Miocene Peach Springs Tuff. Roadcuts in basalt. MP 89.0.

1.0

157.5 Precambrian granite and schist. Older literature would probably assign this to Yavapai Schist. Note numerous dikes and sills. MP 88.0.

2.6

160.1 Valentine; mostly Precambrian granite exposures. Flat-topped buttes capped by Miocene Peach Springs Tuff. MP 85.4.

2.1

162.2 Hackberry; Precambrian outcrops on right (north). Note pediment surface to southwest cut on Precambrian rocks of Peacock Mountains. MP 83.3.

3.3

165.5 Bridge over Hackberry Wash; dark-brown knobs are Precambrian schist. MP 80.0.

2.0

167.5 Entering Hualapai Valley. Southern Grand Wash Cliffs to right (northeast) form edge of Colorado Plateau. Alluvial fans and pediment surfaces are at foot of cliffs. MP 78.0.

2.0

169.5 Good view of Cambrian-Precambrian to north. Muav Limestone is light-colored cliff on skyline. MP 76.0.

1.0

170.5 Road completes "end run" around northern end of Peacock Mountains and turns southwest toward Kingman. MP 75.0.

1.0

171.5 Excellent basin-and-range topography with mature mountain blocks in distance (Cerbat Mountains) and smaller inselbergs near highway. Playa lake beds floor the lower Hualapai Valley 20 miles (32 km) north. MP 74.0.

11.0

182.5 Road on alluvial fan surface that extends from Peacock Mountains to center of Hualapai Valley. MP 63.0.

2.0

184.5 Entering Kingman, Mohave County Seat. Kingman began as a railroad stop on the Santa Fe in the 1880's. It is the largest city in Mojave County and one of the fastest growing in Arizona. MP 61.0.

1.4

185.9 Logasville, trailer park and low-cost housing development on outskirts of Kingman. MP 59.6.

5.3

191.2 Ramada Inn on left. MP 54.3.

END OF FOURTH DAY ROAD LOG

STOP DESCRIPTIONS — FOURTH DAY

Stanley S. Beus,[1] Richard R. Rawson,[1] and Ronald C. Blakey[1]

STOP 1 — CHINO VALLEY OVERLOOK
Near center of T. 22 N., R. 5 W.

We are at the southern edge of the Colorado Plateau. Before you is a view across upper Chino Valley (fig. 9). To the southeast is Picacho Butte, southwest are the Juniper Mountains, due west is Anvil Rock, a prominint butte with a tilted top of Cenozoic basalt capping red beds of the Supai Group. To the west-northwest, just beyond the town of Seligman, is Chino Point, the southern end of a promontory formed by the Aubrey Cliffs (the site of Stop 2). The block "S" northeast of Seligman is on slopes eroded in the Permian Coconino Sandstone.

In 1858 John Newberry, a geologist and member of the Ives Expedition, made one of the first recorded fossil collections from the Permian Kaibab Limestone along the Aubrey Cliffs somewhere north of Chino Point.

STOP 2 — CHINO POINT
West edge of T. 22 N., R. 6 W.

Chino Point, a few miles west of Seligman, Arizona, is the southernmost part of the Aubrey Cliffs. A complete Permian section is present including Esplanade, Hermit, Coconino, Toroweap, and Kaibab Formations (fig. 10). The Schnebly Hill formation is absent but is present 14 miles southeast at Picacho Butte. Chino Point provides one of the best exposures of the basal Esplanade conglomerate and the Permian-Pennsylvanian boundary. At the point where U.S. 66 cuts through the southern edge, Pennsylvanian rocks are exposed to the south of the highway and Permian rocks to the north.

The Pennsylvanian rocks are poorly exposed through a band over one mile wide. Exposures to the south along Interstate 40 consist of cross-stratified sandy oolitic carbonate and quartzose sandstone. Stratigraphic position is uncertain but similar units are present in the Manakacha Formation to the northwest. The railroad cuts south of U.S. 66 expose reddish-brown very limey silty sandstone of the Wescogame Formation. These units are overlain by the lowest of several sedimentary-pebble conglomerates that mark the Permian-Pennsylvanian boundary (figs. 11, 12). The conglomerate clasts range to 4 inches (10 cm) in diameter and are composed of limestone and limey siltstone of various compositions and colors, but all of which can be found as lithologic types in the underlying Pennsylvanian rocks. Sand, silt, dolomite, and calcite spar fill in the voids between the pebbles. The pebbles are rounded and many are elongate but no well-developed examples of imbrication were observed.

The conglomerate occurs as thin, irregular sheets and thicker, lenticular channel deposits. Conglomerate beds are interbedded with sandstone and siltstone typical of other units in the Esplanade Sandstone.

That this conglomerate zone marks the Permian-Pennsylvanian boundary even though fossils are sparse at this and other nearby sections is well-documented by regional evidence. The Wescogame is Virgilian as dated by fusulinids in the western Grand Canyon including some large *Triticites* that will be seen at Iceberg Ridge later in the trip. The age of the Esplanade is based on regional relationships with lower Wolfcampian carbonates. In the western Grand Canyon the Esplanade grades westward into the Pakoon Limestone. Wolfcampian fusulinids are abundant in this formation. At Iceberg Ridge the conglomerate occurs between Virgilian and Wolfcampian fusulinid-bearing strata.

The Esplanade basal conglomerate is succeeded by sandstone, siltstone, and micritic limestone of the main body of the Esplanade Sandstone. Numerous marine fossils seen in thin section in the overlying limey sandstone units attest to the marine affinities of the formation.

Eastward the position of the sub-Permian unconformity can be traced through outcrops in Chino and Verde Valleys to the Midgley Bridge location. Although conglomerate is absent in more locations than it is present, the position of the unconformity can be well established as far east as Sedona. Farther east along the Mogollon Rim the Esplanade pinches out or changes facies, outcrops become more widely spaced, and the exact position of the boundary becomes more uncertain.

STOP 3 — GRAND CANYON CAVERNS
S.E. ¼ T. 25 N., R. 9 W.
Coconino County, Arizona

Large solution caverns in the Redwall Limestone extend from ground level down at least 300 feet (91 m). The largest chamber is about 800 feet (245 m) in length. The caverns were known and used as burials by the Hualapai Indians and perhaps others since prehistoric time. In 1927 a Sante Fe Railroad worker stumbled onto the entrance. Exploration by Los Angeles City College students, members of the Southern California Grotto of the National Speleological Society, in 1960, turned up late Pleistocene ground sloth remains among other things. Beginning in the early 1960's the caverns were developed as a major tourist attraction complete with a 21–story elevator.

The name of the caverns has been changed several times apparently for public relations reasons. The nomenclature began with Coconino and Yampai Caverns (appropriate geographically) and proceeded through Dinosaur Caverns (nothing to do with dinosaurs in either space or time), to Grand Canyon Caverns (inappropriate geographically but probably good advertising).

[1]Department of Geology, Northern Arizona University, Flagstaff, Arizona 86011.

Figure 9. Aerial view west across upper Chino Valley, Yavapai and Coconino Counties, Arizona. Prominent butte on left is Anvil Rock; Interstate 40 left center; Sante Fe Railroad, U.S. 66, and town of Seligman, center; promontory on right is part of Aubrey Cliffs composed of Permian strata; "C" marks Chino Point, site of Stop 2.

We will take the guided elevator and walking tour of the caverns. The caves are entirely dry but afford views of a variety of gypsum and calcite crystal growths.

STOP 4 — DEVONIAN-MISSISSIPPIAN CONTACT NEAR PEACH SPRINGS
S.E. ¼ T. 25 N., R. 11 W.
Mohave County, Arizona

The contact between the Devonian Martin Formation and the Mississippian Redwall Limestone is well displayed in the south (left) roadcut about 7 to 8 m above road level (fig. 13). The basal conglomerate and breccia of the Redwall forms a yellowish-brown weathered zone 1 to 2 m thick that grades upward into medium gray limestone and dolomite of the basal Whitmore Wash Member of the Redwall. The underlying Martin Formation is purple-gray medium-crystalline dolomite at the top and grades downward within 4 m to a thin-bedded medium-gray dolomite near road level. The unconformity at the contact represents most of Famennian (latest Devonian) and Kinderhookian (earliest Mississippian) time. The youngest fossils known in the Martin Formation in the Verde and Chino Valleys to the east are Frasnian brachiopods and conodonts (Beus, 1978). The Whitmore Wash Member of the Redwall is mainly Osagian (Tournaisian) in age based upon foraminiferal microfossils (Mamet and Skipp, 1970). However, the Whitmore Wash was deposited in a sea transgressing from west to east across Arizona and the initial deposits in the west probably formed in latest Kinderhookian time (McKee and Gutschick, 1969, p. 588).

Figure 10. Aerial view of Chino Point and Aubrey Cliffs. Permian stratigraphic units are: E – Esplanade Sandstone; H – Hermit Shale; C – Coconino Sandstone; T – Toroweap Formation; K – Kaibab Limestone.

Figure 11. Roadcut on north side of U.S. 66 at Chino Point, west of Seligman. Light-colored beds marked "C" are basal conglomerates of Esplanade Sandstone (Permian); Wescogame Formation (Pennsylvanian) at road level.

Figure 12. Limestone and siltstone pebble-conglomerate at base of Esplanade Sandstone at Chino Point. This conglomerate is taken as base of Permian.

Figure 13. View of Martin-Redwall unconformable contact in south roadcut at Stop 4. R – Redwall Limestone (Mississippian); M – Martin Formation (Devonian). Unconformity is about at the position of the man's head.

ROAD LOG — FIFTH DAY, ARIZONA PORTION: KINGMAN TO HOOVER DAM, ARIZONA

Stanley S. Beus[1]

Our route will take us from Kingman, Arizona, to Las Vegas, Nevada. We will travel north along the Cerbat Mountains and Grand Wash Cliffs to the upper end of Lake Mead at the Arizona-Nevada border. From South Cove we will travel by boat 9 miles up the lake to Iceberg Ridge where we can traverse a complete Carboniferous stratigraphic section beginning in the Devonian and ending in the Permian. From Iceberg Ridge the route returns to U.S. Highway 93 and then northwest to Las Vegas with a stop at Hoover Dam en route. Total distance of the Arizona portion is 183 miles. Arizona mileposts are noted where available.

The road log and stop descriptions for the fifth day were prepared with the assistance of John J. Matthews, Union Oil Company, and Edwin D. McKee, Ivo Lucchitta, and Ray Douglass of the U.S. Geological Survey.

SEGMENT 1 — Kingman to South Cove.

Total Mileage

0.0 Beginning at milepost 54.4, Ramada Inn, Kingman, proceed west and south on U.S. 66 through the older part of Kingman.

0.1

0.1 Roadcut in Peach Springs Tuff (Miocene) rim that divides "upper" and "lower" Kingman. MP 54.3.

2.3

2.4 Turn right onto U.S. 93, follow signs for Hoover Dam and Las Vegas. Old Santa Fe engine 3759 in Kingman Railroad park on left (west) after turn. Begin new milepost system on U.S. 93 with MP 71.6.

0.6

3.0 Leaving Kingman. Underpass beneath I–40. MP 71.0.

0.8

3.8 Prominent columnar jointing in cliff of Tertiary volcanics on right. MP 70.2.

1.8

5.6 Roadcut in tuffs and agglomerates on right. MP 68.4.

0.4

6.0 Road descends into Sacramento Valley. Black Mountains to northwest (10–12 o'clock) are composed of Precambrian metamorphic and Tertiary (?) volcanic rocks. MP 68.0.

0.9

6.9 Junction with Arizona 68 to Bullhead City on left. Bullhead City, 25 miles (35 km) west on the Colorado River, is the site of a major power generating plant that uses Cretaceous coal from Black Mesa in northeastern Arizona, nearly 300 miles (490 km) away. The coal is delivered through a slurry pipeline at the rate of 300,000 tons per month. MP 67.1.

2.1

9.0 Jagged skyline to left (west) is Tertiary (?) andesite on crest of Black Mountains. Cerbat Mountains to right (east) composed of Precambrian granite and schist. MP 65.0.

2.9

11.9 Historical marker for Cerbat, an abandoned mining town 3 miles northeast. MP 62.1.

3.4

15.3 Junction with Mineral Park Road on right. Road leads to Mineral Park mining property of Duval Corporation (fig. 14); note extensive tailings along mountain front. Earliest mine activity in what is now the Walapai Mining District was for turquoise by Indians. Gold was discovered in 1863 and active mining of gold and silver commenced in the 1870's. Approximately $24,000,000 in gold, silver, lead and zinc was produced through 1946. Copper and molybdenum production began in 1964 and is the main activity today. Mineralization is in Precambrian quartz monzonites. MP 58.7.

1.2

16.5 Santa Claus on left. MP 57.5

4.7

21.2 Grasshopper Junction. Road to right leads to Chloride, a mining town in 1864. MP. 52.8.

4.8

26.0 Volcanic cone remnant on right; road descends into Detrital Valley. MP 48.0.

5.0

31.0 Jagged skyline of Cerbat Mountains to right (east). Tertiary volcanics resting on Precambrian granite and schist. MP 43.0.

1.1

32.1 Junction with Pierce Ferry-Dolan Springs road; turn right toward Pierce Ferry. End of milepost system at MP 41.9.

1.8

33.9 Cerbat Mountains to right front and White Hills to left front composed of Precambrian metamorphic and Tertiary volcanic rocks.

4.2

38.1 Dolan Springs; desert retirement community.

1.1

39.2 North end of Cerbat Mountains on right.

0.7

39.9 Mount Tipton, Precambrian metamorphic rocks capped

[1] Department of Geology, Northern Arizona University, Flagstaff, Arizona 86011

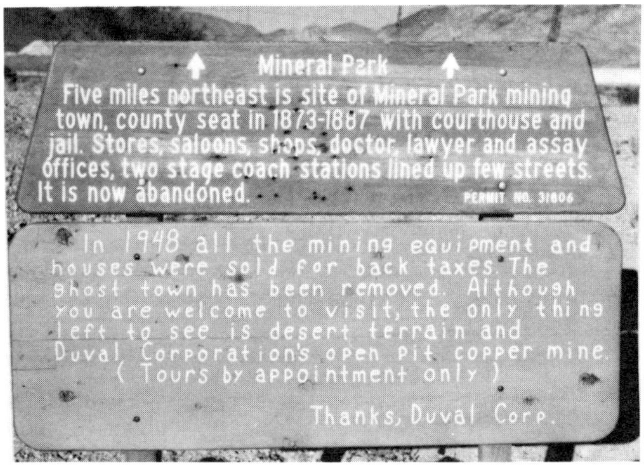

Figure 14. Historical marker signs at turnoff to Mineral Park Mine, milepost 58.7.

by gently tilted Tertiary volcanics at 12 o'clock.

0.5

40.4 Mt. Tipton School.

3.8

44.2 Road on alluvial fan deposits.

3.0

47.2 View across Hualapai Valley to the southern Grand Wash Cliffs which form the western edge of the Colorado Plateau at 12 o'clock.

5.0

52.2 Road descends into northern Hualapai Valley, a closed basin at the eastern edge of the Basin and Range Province.

0.5

52.7 Road junction with Hackberry road on right.

8.5

61.2 **STOP 1 — GRAND WASH CLIFFS;** view of Paleozoic strata, western edge of Colorado Plateau.

7.6

68.8 Boundary of Lake Mead Recreational Area.

1.1

69.9 Junction with road to Meadview, a retirement community, on right. Continue straight ahead on Pierce Ferry Road.

2.2

72.1 Road traversing Grape Vine Mesa on limestone and conglomerates of the Muddy Creek Formation which filled the Grand Wash Trough in Miocene and early Pliocene time.

4.5

76.6 Road on right leads to Pierce Ferry air strip.

0.1

76.7 Road begins descent through Hualapai Limestone Member of the Muddy Creek Formation.

2.2

78.9 Flat-topped cliffs to right are in the Hualapai Limestone Member of the Muddy Creek Formation. Sandstones and conglomerates form slopes below cliff. In our descent of the road to South Cove we cross the Wheeler Fault, a major high-angle fault between Iceberg and Wheeler Ridges having 5,000-10,000 ft (1,520-3,040 m) of vertical movement. Local offset and tilting of the Hualapai Limestone Member of the Muddy Creek by the Wheeler fault indicates recurrent movement in late Pliocene and Pleistocene time.

0.8

79.7 Roadcuts in Mississippian and Pennsylvanian limestones.

0.4

80.1 View to right (west) across Lake Mead into Nevada.

0.8

80.9 Cambrian (?) undifferentiated dolomite in roadcuts.

0.5

81.4 Flat beds of Muddy Creek Formation in roadcuts.

0.6

82.0 South Cove, east shore of Lake Mead. From here we will travel by boat 9 miles north to the northern end of Iceberg Ridge.

9.0

91.0 **STOP 2 — ICEBERG RIDGE.** Traverse of Carboniferous section, discussion of stratigraphy and Pennsylvanian-Permian boundary. Return to South Cove, retrace road route to junction of Pierce Ferry Road and U.S. 93 (49.9 miles).

SEGMENT 2 — Pierce Ferry Road Junction to Hoover Dam.

Mileage begins with Arizona mileposts on U.S. 93 at junction with Pierce Ferry Road.

Total Mileage

0.0 Junction of Pierce Ferry Road with U.S. 93; turn right (north) toward Hoover Dam and Las Vegas. MP 41.9.

2.1

2.1 Texaco Station, "Smith City," on left. MP 39.8.

1.8

3.9 White Hills on right (east) expose a colorful variety of Tertiary volcanics. MP 38.0.

2.1

6.0 Crossing Detrital Wash. MP 35.9.

2.9

8.9 Tertiary volcanics on crest of Black Mountains on left (west). MP 33.0.

8.0

16.9 Precambrian rocks on left (west) in Black Mountains, MP 25.0.

2.0

18.9 Dark knobs on left (west) are Tertiary volcanics at northern end of Black Mountains. MP 23.0.

1.0

19.9 Prominent nipple to northeast at 2 o'clock is Squaw Peak, capped with Tertiary volcanics. MP 22.0.

1.0

20.9 View of Lake Mead to northeast down axis of Detrital Valley; Virgin Mountains in distance. MP 21.0.

1.9

22.8 Junction with road to Temple Bar on right. MP 19.1.

2.1

24.9 Entering Lake Mead Recreational Area. MP 17.0.

1.0

25.9 Roadcuts in Precambrian igneous and metamorphic rocks at south end of Wilson Ridge for next several miles. MP 16.0.

1.8

27.7 Wilson Beach turnoff on left. MP 14.2.

0.4

28.1 Roadcuts are in cemented Tertiary gravels of probable post-Muddy Creek age resting on Precambrian metamorphic rocks. The age and relationship of these cemented gravels to the Muddy Creek Formation is not yet clear. The gravels are related to a through-flowing Colorado River drainage system which is considered a post-Muddy Creek event. However, basalts interbedded with these gravels are indistinguishable from the nearby Fortification Basalt within the upper Muddy Creek Formation that has yielded radiometric dates of 5.5 m.y. or older (Anderson and others, 1972, p. 278). MP 13.8.

1.1

29.2 Viewpoint for Colorado River on left. MP 12.7.

0.7

29.9 Numerous roadcuts through ridges of dissected Tertiary fanglomerates on west flank of Wilson Ridge. MP 12.0.

3.0

32.9 Rest area on right. MP 9.0.

0.5

33.4 Dark-gray 1–m thick camptonite dike containing huge black phenocrysts of kaersutite cuts through fanglomerate in roadcut. Damon et al. (1967) report a minimum date of 3.7 ± 0.7 m.y. for this dike. MP 8.5.

1.8

35.2 Basalt capping fanglomerate in roadcuts. MP 6.7.

0.4

35.6 Isolated lava flow remnants cap small butte on right. MP 6.3.

2.8

38.4 Light-gray and green Tertiary tuff beds beneath dark-gray volcanics in roadcuts. MP 3.5.

0.1

38.5 High-angle fault in volcanics and fanglomerates on right in roadcut. MP 3.4.

0.4

38.9 View to right of basalt flows capping Fortification Hill, high mesa on right. MP 3.0.

2.0

40.9 First viewpoint of Hoover Dam on right. MP 1.0.

1.0

41.9 Cross Hoover Dam, Colorado River, and Arizona-Nevada border.

END OF FIFTH DAY ROAD LOG, ARIZONA PORTION

STOP DESCRIPTIONS — FIFTH DAY, ARIZONA PORTION

Stanley S. Beus[1]

STOP 1 — VIEW OF GRAND WASH CLIFFS
Southern part of T. 29 N., R. 17 W.
Mohave County, Arizona.

Before you is the Grand Wash Trough, filled with late Cenozoic terrestrial deposits and volcanics and bounded on the east by the Grand Wash Cliffs. These cliffs mark the western edge of the Colorado Plateau and the eastern boundary of the Basin and Range Province. They are the result of faulting and erosion along the Grand Wash Fault during late Cenozoic time. The stratigraphic section exposed includes Precambrian metamorphic rocks at the foot of the cliffs through Cambrian, Devonian and Mississippian strata (fig. 15). A few isolated patches of the Pennsylvanian Watahomigi Formation (lower Supai Group) occur along the rim of the cliffs.

The nomenclature applied to Paleozoic strata west of here is generally that of the basin or geosynclinal deposits of Nevada and reflects both real facies changes and local preferences of stratigraphic workers.

STOP 2 — ICEBERG RIDGE
Eastern edge of T. 32 N., R. 17 W.
Mohave County, Arizona

Introduction

Iceberg Ridge is an elongate, eastward-tilted fault block at the eastern edge of the Basin and Range Province and about 5 miles (8 km) west of the western edge of the Colorado Plateau in northwestern Arizona (fig. 16). It is surrounded on three sides by Lake Mead and displays a nearly complete Paleozoic stratigraphic section. The aggregate thickness of Paleozoic strata in this region is about 7,000 feet (2,100 m) and includes rocks of Cambrian, Devonian, and Mississippian-Permian age (fig. 17). Carboniferous strata are magnificently exposed on the northwest face and crest of the ridge. Our traverse will begin in the Devonian Temple Butte Formation at the lakeshore on the west and end at the top of the Pakoon Limestone (Permian) at the lakeshore on the east. Carboniferous strata exposed are the Mississippian Redwall Limestone, 950 feet (290 m) thick, and the Pennsylvanian Callville Limestone, 460 feet (140 m) thick (fig. 18).

The Pennsylvanian-Permian section is particularly instructive because it represents the transition between the mostly clastic facies deposited on the continental shelf to the east (Grand Canyon) and the predominantly carbonate facies of the Cordilleran Miogeosyncline to the west in Nevada. Critical fusulinid horizons permit reasonably accurate correlations between these platform and geosyncline deposits.

[1] Department of Geology, Northern Arizona University, Flagstaff, Arizona 86011.

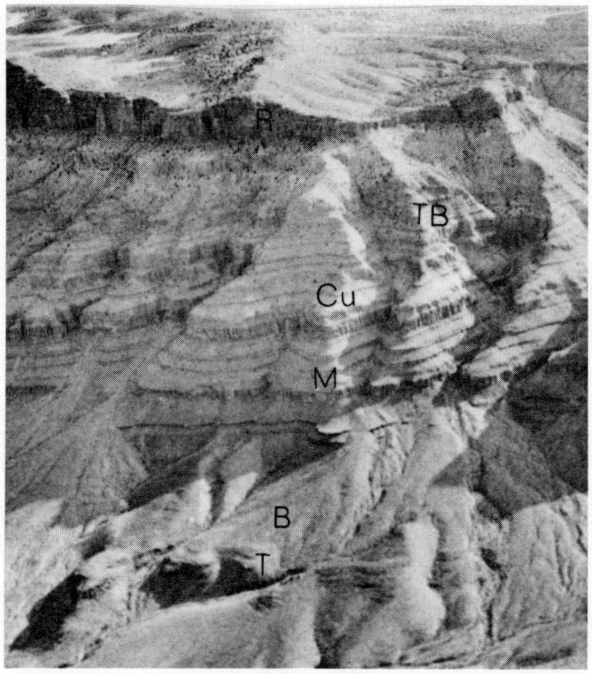

Figure 15. Aerial view of Grand Wash Cliffs, western edge of Colorado Plateau. R – Redwall Limestone (Mississippian); TB – Temple Butte Formation (Devonian); Cu – Cambrian undifferentiated; M – Muav Limestone, B – Bright Angel Shale, T – Tapeats Sandstone (Cambrian).

Redwall Limestone

The Redwall Limestone is well displayed on the west face of Iceberg Ridge from about 200 feet (62 m) above the lakeshore to near the crest of the ridge. The base of the Redwall is generally well marked by a reddish oxidized zone of breccia and conglomerate from 0 to 2 m thick. The upper Temple Butte Formation beneath the Redwall is a medium-gray dolomite in contrast to the lighter gray oolitic limestone of the basal Redwall.

The Whitmore Wash Member of the Redwall is a light-gray oolitic limestone in the lower part and includes several stromatolite beds (units 1 and 2, fig. 18). The upper part is a light-gray to white crystalline dolomite. The Thunder Springs Member, the most distinctive in the Redwall, is characterized by abundant stringers, nodules, and thin beds of light-gray to pink chert 2 to 15 cm thick, interbedded with dark-gray micritic limestone. The chert weathers red brown and imparts an overall dark-brown color to the Thunder Springs Member which outcrops as a dark stripe in the otherwise light-gray cliff of the Redwall (see fig. 19).

Figure 16. Aerial view to northeast of Iceberg Canyon, upper end of Lake Mead. Azure Ridge in Nevada, on left, and Iceberg Ridge in Arizona, on right. Similar sections of steeply dipping Paleozoic strata occur on each ridge.

The Mooney Falls Member is unusually thick (414 ft (125 m)) and is mainly a light-gray to white crystalline dolomite which has replaced what was originally a crinoidal limestone. This is in contrast to the outcrops we have seen in eastern Grand Canyon and the Verde Valley where the Mooney Falls is generally the least dolomitized of the Redwall members. (Rawson and Kent, this guidebook)

The contact of the Mooney Falls with the overlying Horseshoe Mesa Member is somewhat obscure. In this section it is placed at the first appearance of very thin-bedded aphanitic dolomite and limestone that is typical of the Horseshoe Mesa. The Horseshoe Mesa Member is light-gray thin-bedded limestone and dolomite with some chert beds and forms generally less resistant and less massive-appearing ledges and slopes than the Mooney Falls Member below. The contact of the Horseshoe Mesa Member with the Callville Limestone above (Mississippian-Pennsylvanian boundary) is largely obscured by rubble. The boundary is placed near the bottom of a small ravine where thin-bedded cherty purplish to pinkish-gray limestone beds are first encountered above the light-gray limestone of the upper Redwall.

Megafossils are abundant in the Redwall Limestone (see fig. 18), particularly in the Thunder Springs Member, and include corals, brachiopods, bryozoans, crinoids, gastropods, and bivalves, many of which are described in McKee and Gutschick (1969). A Kinderhookian-Meramecian age is assigned to the Redwall in northwestern Arizona by McKee and Gutschick (1969), based upon coral and foraminiferal zones related to those in the midcontinent (see also Mamet and Skipp, 1970).

Pennsylvanian–Permian Strata

The Pennsylvanian Callville Limestone overlies the Redwall and outcrops on the crest and upper east flank of Iceberg Ridge (figs. 19, 20). It is characterized by thin-bedded gray limestone and red-brown sandstone and siltstone and forms a series of irregular ledges and slopes. Wedge-planar and tabular-planar cross-bedding is common through much of the Callville. The thickness of the Callville, 460 feet (140 m), is somewhat less than that recorded at Iceberg Canyon (238 m) a few miles to the north-northwest by Lumsden et al. (1973), and elsewhere on Iceberg Ridge (761 ft (231 m)) by Edwin D. McKee (1978, personal commun.). The differences in thickness may be in part due to the imprecise placement of the lower boundary. McKee (1978, personal commun.) has recognized the lower three Supai Group formations or equivalents on Iceberg Ridge within the Callville Limestone. In this section (fig. 18) units 1-8 appear equivalent to the Watahomigi Formation, units 9-11 to the Manakacha, and units 12-14 to the Wescogame Formation of McKee (1975).

Fossils are abundant throughout much of the Callville section and fusulinid foraminifers (identified by Ray Douglass) provide significant dateable horizons in this section. *Triticites plummeri* in unit 12, just over the crest of the ridge and near the top of the Callville, indicates a Virgilian age (Ray Douglass, 1978, personal commun.) and Permian fusulinids including *Schwagerina* sp. mark an important Wolfcampian horizon in unit 16, the lower part of the Pakoon Formation. The age assignment of Morrowan-Desmoinesian for units 1-11 of the Callville is based upon the age of laterally equivalent foraminifer-bearing beds in the Supai Group in Grand Canyon (McKee, 1975; 1978, personal commun.).

The Pennsylvanian-Permian boundary, the contact between the Callville and Pakoon Limestones, is placed at the base of a thin and irregular limestone pebble conglomerate, unit 15 (fig. 21), which appears to be identical with the conglomerate used by Mckee (1975; 1978, personal commun.) to mark the Pennsylvanian-Permian boundary (Wescogame-Esplanade contact) in western Grand Canyon. Early Permian (Wolfcampian) fusulinids occur in gray limestone beds of unit 20, about 60 feet above the boundary.

The latter part of our traverse will pass down dip through the Pakoon Limestone to the lakeshore on the east side of Iceberg Ridge. The Pakoon is about 250 feet (76 m) thick and consists mainly of dolomitic limestone with some sandstone and mudstone. Fossils are abundant in the limestone beds and include solitary and colonial corals, brachiopods, fusulinids, gastropods, bivalves, crinoids, and echinoids.

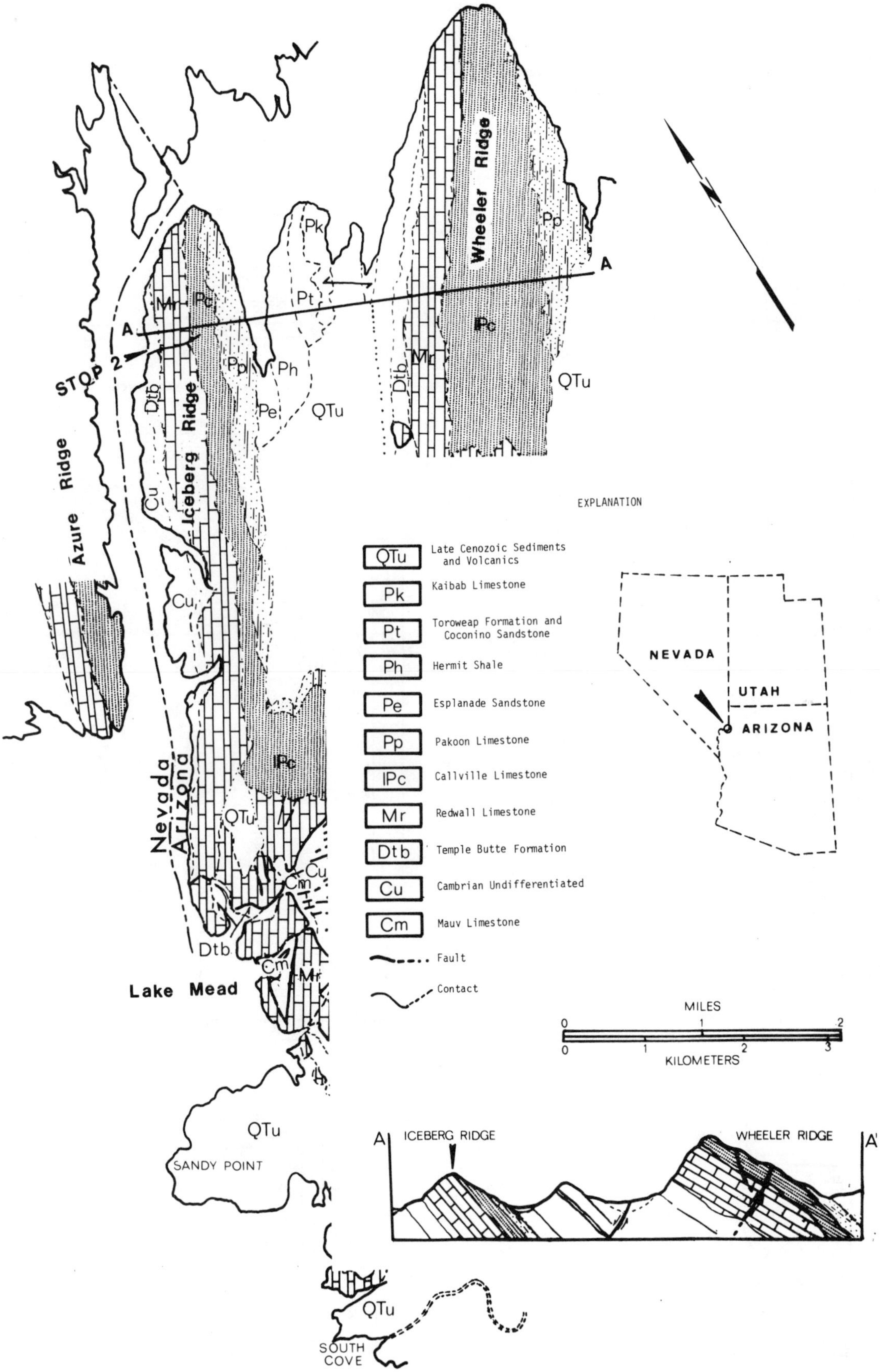

Figure 17. Geologic map of Iceberg and Wheeler Ridges at edge of Lake Mead, Mohave County, Arizona (modified from Matthews, 1976).

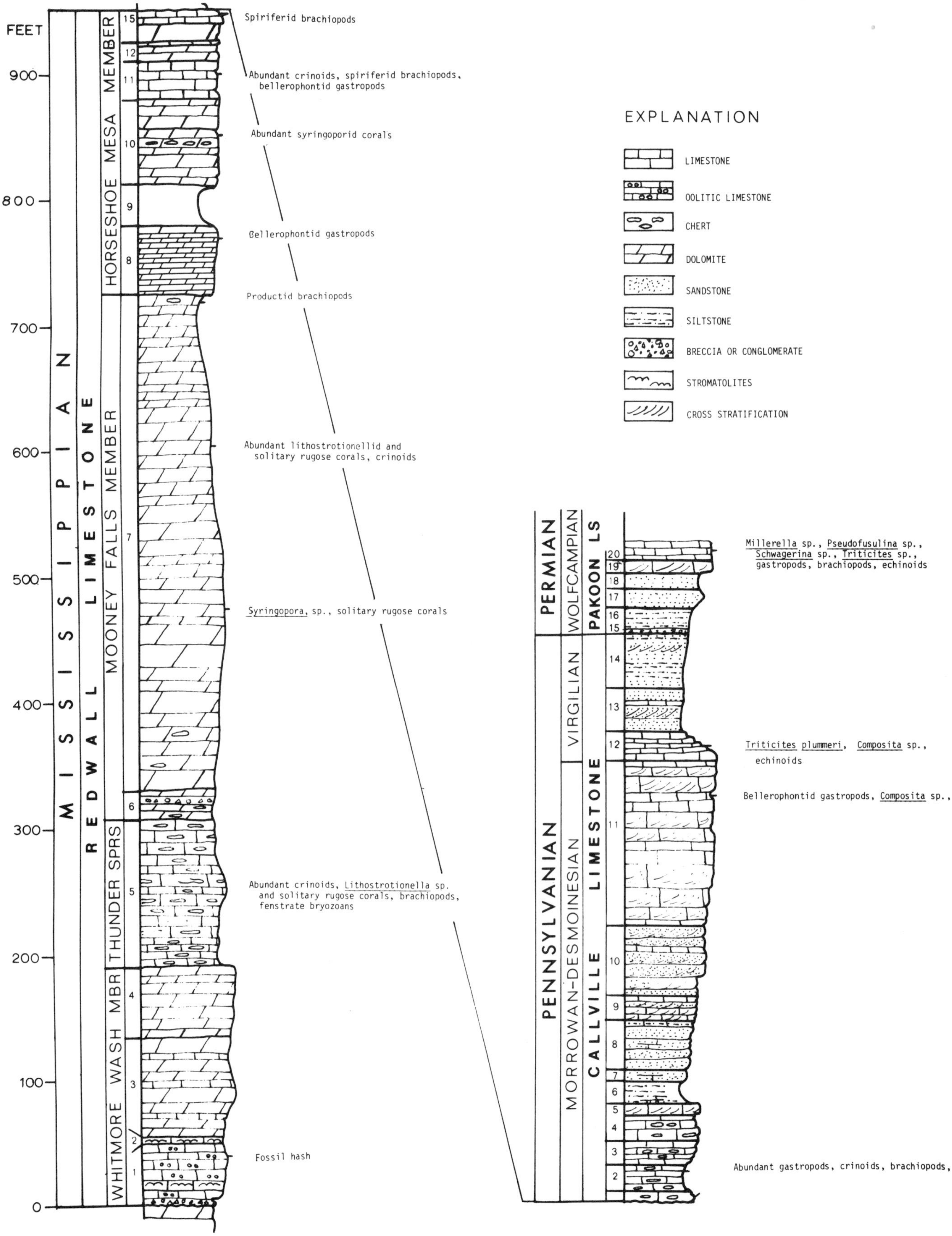

Figure 18. Carboniferous stratigraphic section exposed along traverse over Iceberg Ridge.

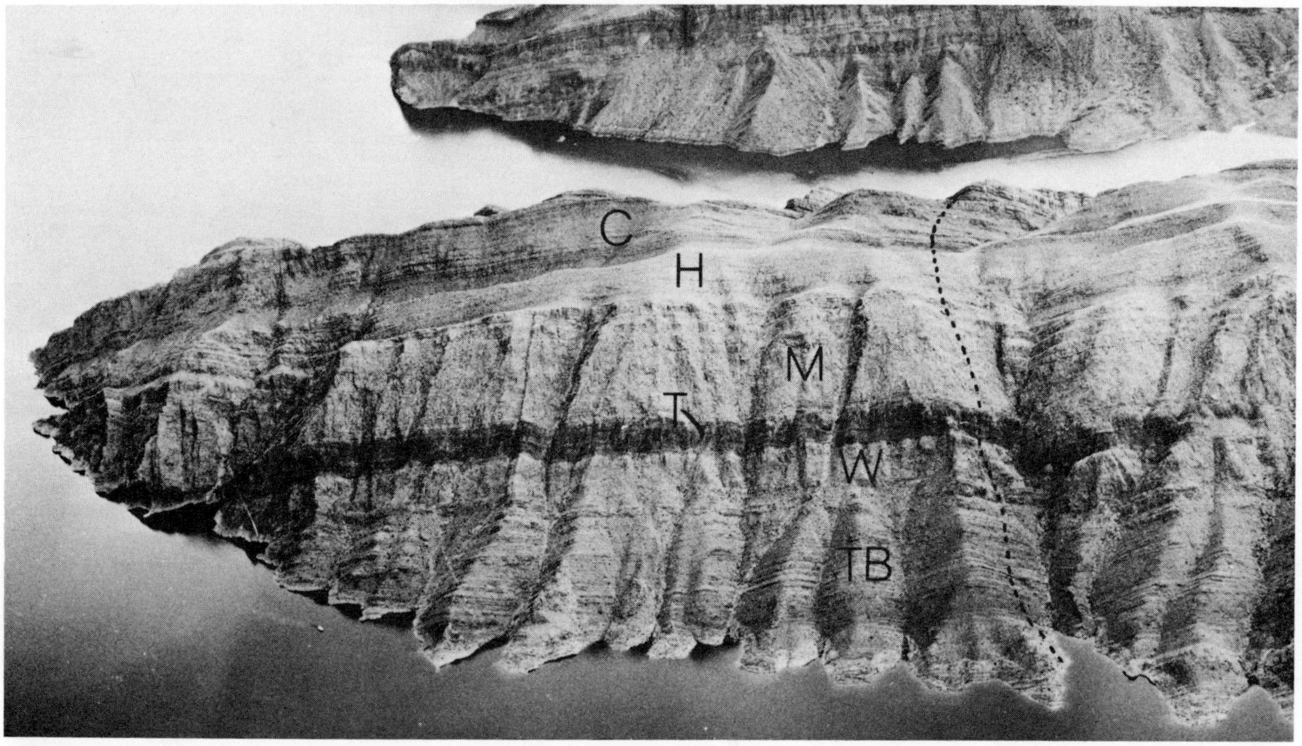

Figure 19. Aerial view of west face of Iceberg Ridge, east shoreline of Lake Mead. Dashed line shows route of traverse at Stop 2. C – Callville Limestone (Pennsylvanian); Redwall Limestone (Mississippian): H – Horseshoe Mesa Member, M – Mooney Falls Member, T – Thunder Springs Member, W – Whitemore Wash Member; TB – Temple Butte Formation (Devonian).

Figure 20. Aerial view to southwest of Iceberg Ridge. R – Redwall Limestone (Mississippian); C – Callville Limestone (Pennsylvanian); P – Pakoon Limestone (Permian).

Figure 21. View of steeply east-dipping beds of Pakoon and Callville Limestones high on east flank of Iceberg Ridge. Arrow indicates approximate position of limestone-pebble conglomerate that marks base of Pakoon Limestone and presumed Pennsylvanian-Permian boundary.

ROAD LOG — FIFTH DAY, NEVADA PORTION: HOOVER DAM TO LAS VEGAS, NEVADA

Ralph L. Langenheim, Jr.[1] **and Gary D. Webster**[2]

The field trip enters Nevada, crossing Hoover Dam, via U.S. 93 and proceeds northwesterly to Las Vegas by way of Boulder City and Henderson, Nevada (figs. 22,23). The road log ends at the junction or divergence of U.S. 95 and 93 at the intersection of Fremont Street and Las Vegas Boulevard in downtown Las Vegas. Should you wish to proceed directly to stops on the "Sixth Day: Clark County, Nevada" log, leave this route at mile 25.7, turn right and proceed northward on Nellis Boulevard to its junction with Las Vegas Boulevard. This intersection is mile 5.8 on the "Sixth Day: Clark County, Nevada" log. Should you wish to proceed directly to the "Strip," turn left at mile 28.9 on Sahara Boulevard and proceed directly west to intersect the northern end of the Strip on Las Vegas Boulevard. The Strip also may be reached via Tropicana Boulevard, Flamingo Road, and Desert Inn Road, all of which diverge to the left from U.S. 93–95.

Total Mileage

0.0 Log begins at entrance to Visitors Center at north end of Hoover Dam and just before passing monument. Rock here is latitic flow breccia and tuff probably belonging to Golden Door Volcanics of Longwell (1963). These rocks probably are middle to late Miocene (Anderson et al., 1972).

0.1

0.1 Parking lot and viewpoint to left.

STOP 1 — HOOVER DAM. View downstream face of dam, powerhouses, Black Canyon and Tertiary volcanic rocks.

0.3

0.4 Good view of Fortification Hill at 12 o'clock. Fortification Hill, located on Arizona side of reservoir, is composed of Miocene–Pliocene Muddy Creek Formation resting on older Tertiary and undifferentiated volcanic and intrusive rocks. Basalt capping mesa is type locality of Fortification Basalt Member of Muddy Creek Formation by Longwell (1963).

0.2

0.6 Transformer station. Bedrock is Golden Door Volcanics.

0.4

1.0 Pullout space, a good place to get a close look at Golden Door Volcanics.

0.7

1.7 Turn off to right to enter Lakeview Viewpoint.

0.2

1.9 Parking lot.

STOP 2 — LAKEVIEW VIEWPOINT. Panoramic view of lower Lake Mead Reservoir and bordering mountain ranges. Return to U.S. 93.

0.2

2.1 Turn right toward Boulder City and Las Vegas. Latitic tuff and tuff breccia of Golden Door Volcanics extensively sheared in roadcut outcrops and slopes north of highway ascending hill. Tuff appears darker colored and more dense. Some of the tuff breccia is very coarse grained. Highway nearly parallels contact between Boulder City Pluton to left and Golden Door Volcanics to right.

1.6

3.7 Gold Strike Casino, gasoline station, commercial museum and helicopter pad to right. Granodiorite in Boulder City Pluton crops out along ridgeline to left at 9 o'clock.

0.8

4.5 Lake Mead Information Center to right. Nevada 41 descends toward Lake Mead, giving access to lakeshore, a public campground (with showers), and a marina. The Information Center contains a small natural history collection, toilet facilities, and a limited stock of books and pamphlets for sale.

0.2

4.7 Pullout to right gives an opportunity to take pictures and/or appreciate the view. Saddle Island, at 2:30, is composed of Precambrian schist. At low water levels, Saddle Island is connected to shore.

0.4

5.1 Pullout to right. River Mountains, directly ahead at 12 o'clock, are composed mostly of Golden Door Volcanics, but mass directly ahead is intruded by an eruption center flanked by a small area of early to middle Miocene Patsy Mine Volcanics. Details of this area are not completely resolved.

1.0

6.1 Boulder City water tank at 12 o'clock, is built on a hill of intrusive granodiorite belonging to Boulder City Pluton. All rounded hills to left of road are in the pluton.

0.8

6.9 Villa Drive enters highway from right; continue straight ahead.

0.1

7.0 Bear right, leaving highway to bypass Boulder City on truck route.

0.3

7.3 Roadside rest with picnic tables, but no sanitary facilities, to left.

1.2

8.5 Bear right to rejoin U.S. 93 en route to Henderson and

[1]Department of Geology, University of Illinois, Urbana, Illinois 61801
[2]Department of Geology, Washington State University, Pullman, Washington 99163

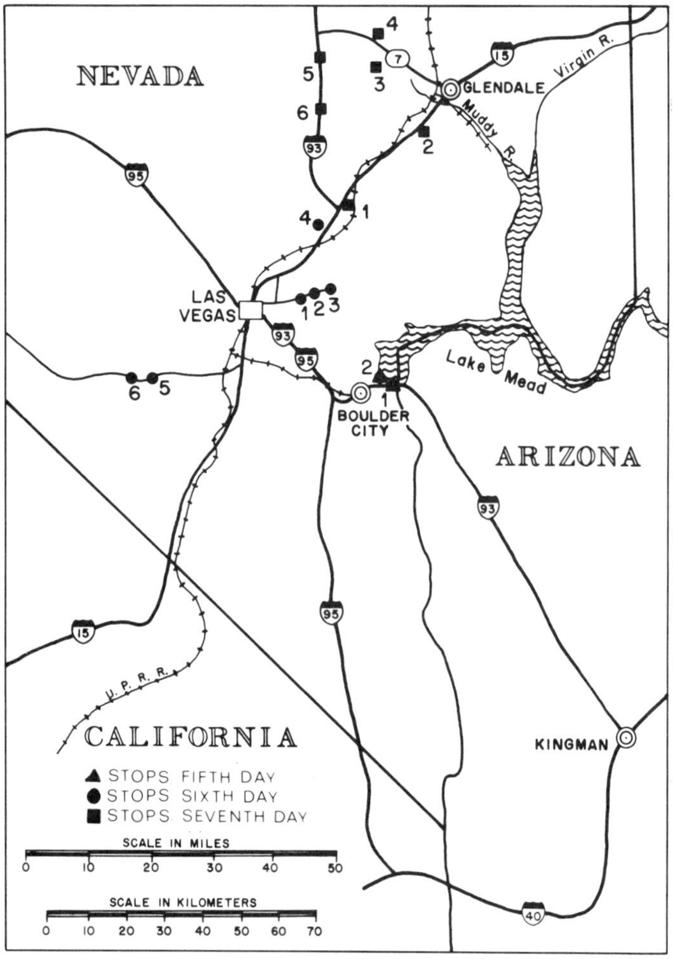

Figure 22. Field trip itineraries: Fifth Day, Hoover Dam – Las Vegas; Sixth Day, Clark County, Nevada; and Seventh Day, Clark County, Nevada.

Las Vegas.

0.1

8.6 Merge with U.S. 93, westbound.

0.7

9.3 Beginning of divided highway. Outlier of McCullough Range at 12 o'clock, River Range to right at 1 to 4:30, northwest end of Opal Mountains at 9 o'clock beyond an extensive alluvial fan, main body of McCullough Range at 10–11 o'clock. Most of River Range visible from this point is composed of Golden Door Volcanics. Opal Range and connected portions of Eldorado Mountains are of Patsy Mine Volcanics and Mount Davis Volcanics respectively. Volcanic rocks composing most of the McCullough Range are so far undivided, but visible portion or outlier at 12 o'clock is part of an intrusive quartz monzonite porphyry.

1.8

11.1 Exit for U.S. 95 southbound to right; proceed straight ahead on U.S. 93–95 northbound.

1.4

12.5 Railroad Pass Casino to right. Mine or prospect on north side of hill behind casino is in a complex mixture of extrusive and intrusive rocks, chiefly andesite and latite, which have been alunitized and silicified. The prospector probably sought gold and/or silver in quartz veins cutting these rocks. This prospect, as well as more promising mines in the Alunite (Railroad Pass) Mining District have not proven lucrative (Longwell et al., 1965). These rocks are mapped by Longwell (1963) as a part of the Golden Door Volcanics.

0.6

13.1 Railroad crossing. Union Pacific Railroad, Boulder City Branch Line.

0.4

13.5 Power line crossing road at crest of Railroad Pass. Railroad Pass is gap between McCullough Range and River Mountains, separating Las Vegas Valley at 12 o'clock from Searchlight Playa at 6 o'clock. Las Vegas Valley drains externally, via Las Vegas Wash, to Colorado River and Gulf of California. Searchlight Playa occupies a local basin of internal drainage. Gass Peak is visible at 12:30 with Sheep Range rising behind it. Frenchman Mountain is a bit closer at 1 o'clock and will be visited tomorrow (Stops 1–3). River Mountains are at 2–3 o'clock. Spring Mountains are distant, high range at 11 o'clock. All of these ranges, except McCullough Range, are composed mostly of Paleozoic and Mesozoic sedimentary rocks ranging in age from Cambrian through Triassic or Early Jurassic, contrasting with mountains of Tertiary volcanic rocks and Precambrian crystalline rocks to southeast. Tall buildings scattered at left of 11 o'clock on valley floor are on Las Vegas Strip; more concentrated cluster at 11–12 o'clock are hotels and casinos of downtown Las Vegas on Fremont Street or "Glitter Gulch." The Nevada Test Site — old A.E.C. Proving Grounds — lies to north on horizon in Las Vegas Valley.

0.4

13.9 Continue northward between McCullough and River Mountains. Low ridges now visible to right of Frenchman Mountain are composed of intricately faulted late Paleozoic, Mesozoic and early Tertiary sedimentary rocks.

2.9

16.8 Muddy Peak has now become visible beyond north end of River Mountains at 2:30–3:00. Sharply pointed peak at 2 o'clock is an isolated fault block of Tertiary volcanic rock.

1.0

17.8 Stoplight at Major Ave. — a principal street in Henderson, Nevada. Muddy Mountains now well displayed on horizon at 3 o'clock.

1.0

18.8 Stoplight at Lake Mead Drive.

0.5

19.3 Former Basic Magnesium plant at 9 o'clock. The chemical industry here began to exploit "cheap" electric power from Boulder Dam to produce magnesium metal for aircraft manufacture during World War II.

3.7

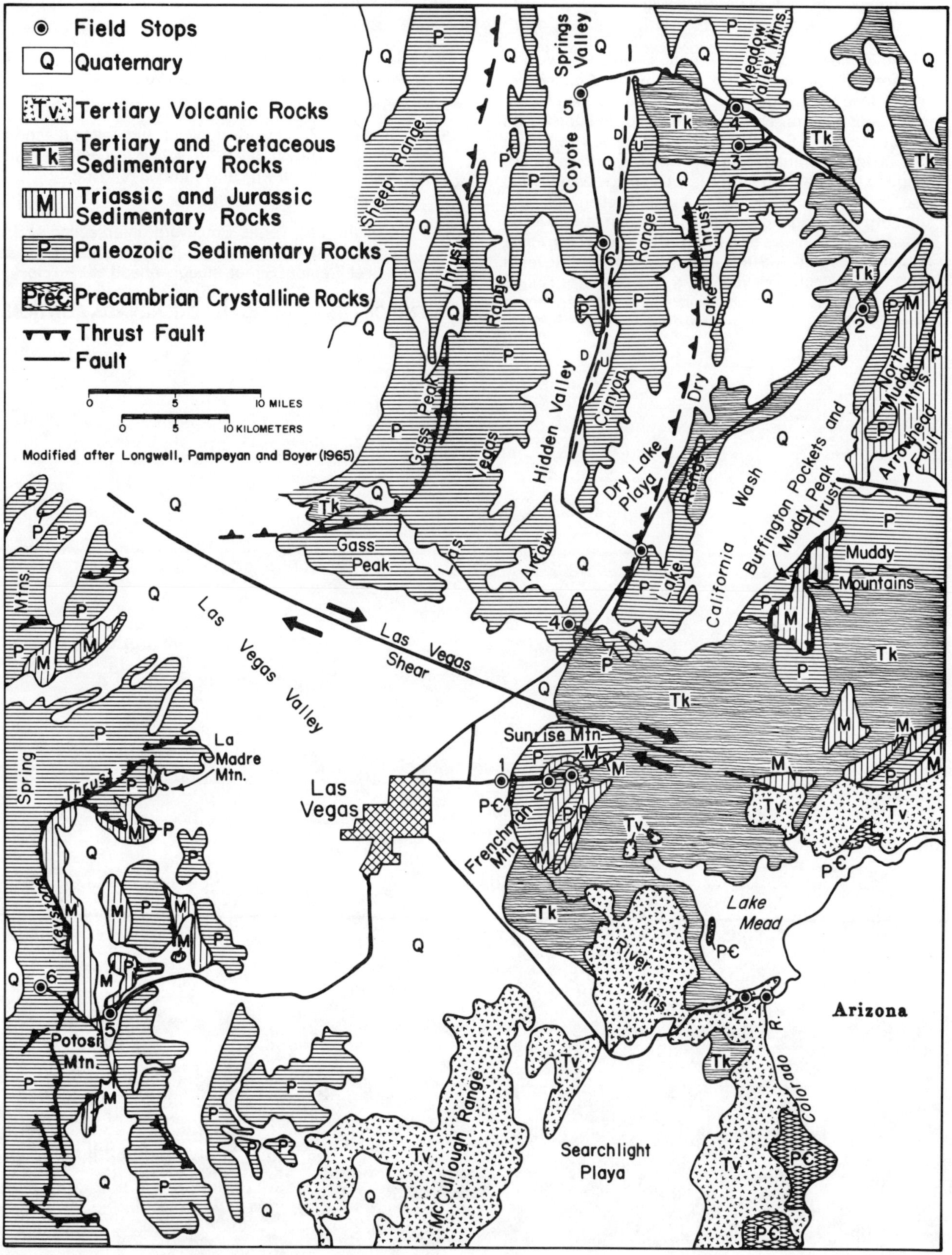

Figure 23. Geologic map of part of Clark County, Nevada, showing trip stops and routes. Geology modified after Longwell et al. (1965).

23.0 Russell Road; power plant to left is oil fired. Nevada's share of Hoover Dam power is no longer adequate to keep the bright lights of Las Vegas lit!

2.7

25.7 Nellis Boulevard diverges to right. Should you wish to proceed directly to the "Sixth Day: Clark County, Nevada" road log and geologic sites on Frenchman Mountain, bear right on Nellis Boulevard to its intersection with Lake Mead Boulevard. This point is at mile 5.8 on the log to Frenchman Mountain. Should you wish to continue to the "Seventh Day: Clark County, Nevada" road log, follow Nellis Boulevard to its intersection with U.S. 93, turn right and proceed to its interchange with Interstate 15. This point is mile 27.6 on the "Sixth Day: Clark County, Nevada" log and mile 16.5 on the "Seventh Day: Clark County, Nevada" log. The Dry Lake Range and Las Vegas Range now visible to left of Frenchman Mountain.

1.6

27.3 Lamb Boulevard, stoplight. Lamb Boulevard may be taken to intersect I–15 northeast of Las Vegas, avoiding much traffic.

1.5

28.8 Stoplight, Sahara Boulevard. Turn left to intersect Strip without entering downtown Las Vegas.

1.2

30.0 Charleston Boulevard, stoplight. Turn left to reach Strip, turn right to reach Salt Lake highways, continue straight ahead for downtown Las Vegas via road log.

1.5

31.5 Stoplight, Las Vegas Boulevard, intersection U.S. 91 and U.S. 93–95. "Glitter Gulch," the downtown segment of Fremont Street, straight ahead at 12 o'clock.

END OF FIFTH DAY ROAD LOG, NEVADA PORTION

STOP DESCRIPTIONS — FIFTH DAY, NEVADA PORTION

Ralph L. Langenheim, Jr.[1] and Gary D. Webster[2]

STOP 1 — HOOVER DAM
NE¼ SE¼ sec. 29, T. 22 S., R. 65 E, Clark County, Nevada

Our stop at the viewpoint at the Nevada end of Hoover Dam provides a view of the dam and some of its downstream supporting structures and an opportunity to quickly identify and characterize the rock units exposed along our trip into Las Vegas.

Further details regarding the dam may be sought from the pamphlet (U.S. Bureau of Reclamation, 1971) supplied by the Federal Government explaining the dam to the public and in Kleinsorge (1941) and Woodbury (1941).

Bedrock in the environs of Hoover Dam is restricted to Precambrian crystalline rocks, Tertiary volcanic and plutonic rocks, and the late Tertiary Muddy Creek Formation (Longwell, 1936, 1963; Longwell et al., 1965; and Anderson, 1969). The thick sequences of Paleozoic and Mesozoic sedimentary rock that characterize the platform sequence in the Grand Canyon and the much thicker miogeosynclinal sequence to the north in Clark County are completely absent here and to the south.

Precambrian crystalline rocks of diverse character crop out in the cores of the mountain ranges paralleling the Colorado River and U.S. Highway 93 between the latitude of Kingman and Lake Mead. In this region, such outcrops in the northern end of the Black Mountains east of the Boulder Basin portion of Lake Mead and in the core of the Eldorado Mountains, starting about six miles south of Hoover Dam, are extensive. In addition, Saddle Island, in the western part of Boulder Basin, is composed of Precambrian rock and there is a small outcropping of Precambrian material at the western foot of Frenchman Mountain at Las Vegas (fig. 22). These rocks generally have been little studied and are described only in reconnaissance fashion. According to Longwell (1963) the most conspicuous rock type is strongly banded gneiss, much of it granite gneiss. Micaceous schist and chloritic schist also are cited as important and the foliation is described as steeply dipping and generally striking northerly.

Rocks visible from the viewpoint are mapped by Longwell (1963) as part of the Eocene or Oligocene (?) Golden Door Volcanics. Anderson (1969), however, suggests that elements of the older Patsy Mine Volcanics also may be included in the Golden Door Volcanics as mapped by Longwell (1963) between Boulder City and Hoover Dam. In a later work, Anderson (1971) recommends abandonment of the Golden Door Volcanics as a stratigraphic unit because the type area is in rock he considers characteristic of the Patsy Mine Volcanics. Although Anderson (1971) proposed an informal unit, the "Tuff of Bridge Spring," as a substitute for "Golden Door Volcanics," we will continue to use Longwell's (1963) terminology for the Boulder City–Hoover Dam region. Anderson et al. (1972), on the basis of two isotopic determinations, indicate a middle to late Miocene, Barstovian, age for the Golden Door Volcanics ("Tuff of Bridge Springs") in the Eldorado area to the south. The Golden Door Volcanics are a complex of latitic to rhyolitic rock with much glass and tuff. They are intruded by dikes, stocks, and irregular bodies of similar character. They are described as being lighter colored than the dark-gray, brown, or black Mount Davis Volcanics, which succeed them. They also are described as lighter colored than the "somber brown" Patsy Mine Volcanics, which are older. The rocks in the road cutting northeast of the viewpoint appear to be latitic crystal tuff.

The Patsy Mine Volcanics, also considered Eocene or Oligocene by Longwell (1963), underlie the Golden Door Volcanics, resting directly on Precambrian rock in the Eldorado Range about six miles south of the dam. Anderson et al. (1972), on the basis of 11 isotopic age determinations, consider the Patsy Mine Volcanics most probably early to middle Miocene or Hemingfordian. Patsy Mine Volcanics are not readily viewed from our traverse, but do crop out in Black Canyon, about one-half mile below Hoover Dam. Longwell (1963) describes these rocks as differing from the Golden Door Volcanics in being predominantly composed of lavas with some thick units of explosion breccia.

The Mount Davis Volcanics, questionably of Miocene age according to Longwell (1963), are placed in the late Barstovian by Anderson et al. (1972) on the basis of 11 radiogenic age determinations. The Mount Davis Volcanics rest unconformably on all older rocks. These volcanics are predominantly basaltic to andesitic flows and agglomerate with interbedded coarse sediments. They are not exposed adjacent to the line of our traverse today, but U.S. 93 does cross about two miles of outcropping Mount Davis Volcanics in Arizona just south of the crest of the grade east of Hoover Dam.

The Boulder City Pluton, which intrudes Patsy Mine and Golden Door Volcanics and is unconformably succeeded by Mount Davis Volcanics, crops out over about ten square miles between Hoover Dam and Boulder City. According to Anderson (1969) the pluton is a composite mass of quartz monzonitic to granitic composition, is predominantly granitoid, and is slightly porphyritic. It is marked by a persistent brown-stained, cliff-forming, nearly horizontal zone within the granite at an elevation of about 2,400 feet (730 m). This zone, previously mapped as the base of a laccolithic or sill-like intrusion (Longwell, 1963), is interpreted by Anderson (1969) as being caused by precipitation of iron oxides at or near the water table about 2.6 to 10.6 m.y. B.P.

Rocks of the Muddy Creek Formation, which include claystone, sandstone, siltstone, conglomerate, salt, gypsum, limestone, and associated basalt, unconformably succeed all older rocks and were deposited in basins formed during Basin and Range faulting. Fine-grained detrital rocks predominate, but the Fortification Basalt Member, a prominent sequence dominated by flow rocks, serves to roughly divide the Muddy Creek Formation into three members. The Fortification Basalt Member, which forms the capping on Fortification Hill east of Boulder Basin, overlaps the lower member, member 1 of Longwell (1963), and rests on rocks of all ages down to and including

[1] Department of Geology, University of Illinois, Urbana, Illinois 61801.
[2] Department of Geology, Washington State University, Pullman, Washington 99163.

Figure 24. **Northwesterly view from Lakeview Viewpoint. 1 – River Mountains; 2 – Dry Lake Range; 9 – unnamed islet south of Saddle Island; 12 – Boulder Beach.**

Figure 25. **North-northwest view from Lakeview Viewpoint; adjoins Figure 23 to left. 2 – Dry Lake Range; 3 – Arrow Canyon Range; 4 – Sheep Range and Las Vegas Ranges; 8 – Saddle Island; 9 – unnamed islet south of Saddle Island; 10 – Boulder Island.**

the Precambrian. Fossils are notably absent in the Muddy Creek Formation and those found so far are ambiguous as to the age of the rocks. The late Miocene to Pliocene (?) age of the unit has largely been inferred from its similarity in lithology and stratigraphic position to the Pliocene Panaca Formation of the Pioche area, about 150 miles north, and to a sequence of Pliocene lake beds in Big Sandy Valley, Arizona (Longwell et al., 1965). Isotopic age determinations, ten by Anderson et al. (1972) and two by Blair, McKee, and Armstrong (1977), however, range from 5.8 to 11.1 m.y. and are consistently Pliocene in age. Furthermore, the distribution of basalts of differing ages within that time range suggests that the unit is not a single, area-wide sheet type eruption. At any rate most of the Muddy Creek Formation thus appears to be definitely Pliocene in age, although some of the material of Longwell's (1936) member 1 may be of late Miocene age.

Until recently, the Muddy Creek Formation was considered to have originated as basin fill in terrestrial environments. Geochemical and sedimentologic analysis of the Hualapai Limestone Member, exposed near the mouth of the Grand Canyon, however, now suggests that at least some of the Muddy Creek Formation was deposited in the head of an ancestral Gulf of California (Blair, 1978).

Basin and Range faults trending northerly to northwesterly dominate the structure of the area immediately around Hoover Dam and Boulder City. These faults generally dip about 45° or less to the west and separate eastward-tilted blocks. Removal of most of the valley fill by erosion during the last 2.6 m.y. (Blair, McKee, and Armstrong, 1977) displays the faulting in much more detail than will be available in the areas to be visited tomorrow and the next day where interior drainage still persists or where downcutting has not completely exhumed the bedrock blocks. In this area we are southeast of the platform-geosynclinal margin and the coincident thrust faulting. In addition, the Tertiary rocks of the region are all younger than the Mesozoic tectonic events responsible for the thrusts and attendant folds in the geosyncline.

STOP 2 — LAKEVIEW VIEWPOINT
SE¼ NE¼, sec. 30,T. 22 S., R. 65 E.
Clark County, Nevada

Mountains visible in the panorama from Lakeview Viewpoint are as follows, from left to right: River Mountains (fig. 24), Saddle Island (with the Dry Lake, Arrow Canyon, Las Vegas and Sheep Ranges beyond) (fig. 25), the Muddy Mountains (figs. 26, 27), the Black Mountains (figs. 26, 27), and the lower slopes of Fortification Hill (fig. 27). The body of water is Boulder Basin, the lower part of Lake Mead. The small islands include the Boulder Islands (figs. 25, 26) and Sentinal Island (figs. 26, 27) as well as an unnamed islet to the left of Saddle Island (figs. 24, 25). Boulder Beach and other developments are located in Hemenway Wash in the foreground to the left (fig. 24). The River Mountains, as seen from this point, are composed of Eocene to Oligocene (?) Golden Door Volcanics. Saddle Island is composed of Precambrian schist. That part of the Dry Lake, Arrow Canyon, and Las Vegas Ranges, barely visible from this point, are mostly composed of Devonian and Carboniferous carbonate rocks. The Sheep Range, as seen from here, is composed of Cambrian and Ordovician sedimentary rocks. The Muddy Mountains, from this aspect, are mostly Paleozoic carbonate rocks, with the topmost peak, Muddy Peak, of Devonian Muddy Peak Limestone. To the right and below Muddy Peak the bright red outcrop is of Triassic Chinle and Moenkopi Formations to the north of a pink outcrop of Aztec Sandstone (latest Triassic or Early Jurassic) and to the south of Permian Kaibab and Pakoon Formations. The Black Mountains, as seen from this point, chiefly expose Tertiary volcanic rocks. The low-lying black outcrops in front of the Muddy Mountains and to their left are chiefly Tertiary volcanic rock (fig. 26) and the lower, lighter colored slopes between the Muddy Mountains and the River Mountains and in front of the Dry Lake Range are mostly an extensive exposure of Muddy Creek Formation.

At this stop we have our first chance to view some of the diverse stratigraphic and structurally distinct areas to be traversed during the remainder of the trip. Perhaps most important from our point of view is the contrast between the stratigraphy of the late Precambrian through early Mesozoic Cordilleran Miogeosyncline and that of the contemporaneous platform. The miogeosyncline was characterized by the deposition of thick, relatively continuous sequences that are dominated by carbonate rocks in the Middle Cambrian through Permian portion of the section in contrast to a more mixed carbonate and detrital succession on the platform. These rocks are more than 37,000 feet (11,250 m) thick in the North Spring Mountains in contrast to only about 11,000 feet (3,350 m) at Frenchman Mountain (Longwell et al., 1965). The platform sequence is present in the Muddy Mountains (figs. 26–28) and the miogeosynclinal sequence is in the dimly visible Dry Lake, Arrow Canyon, and Sheep Ranges (figs. 25, 28) as seen from this point.

Figure 26. View north from Lakeview Viewpoint; adjoins Figure 24. 5 – Muddy Mountains; 6 – Black Mountains; 10 – Boulder Islands; 11 – Sentinel Island.

Figure 27. North-northeast view from Lakeview Viewpoint; adjoins Figure 25. 5 – Muddy Mountains; 6 – Black Mountains; 7 – north slope of Fortification Hill; 11 – Sentinel Island.

Early Carboniferous or Mississippian rocks in this area do not differ greatly in facies between the miogeosyncline and platform, and the geosynclinal Monte Cristo Group is readily correlated in some detail with the Redwall Limestone on the platform. The thickness, however, increases substantially with 800 to 900 feet (240–275 m) of Mississippian rock in Frenchman Mountain and at Mountain Spring Pass in contrast to about 1,600 feet (480 m) in the more geosynclinal Arrow Canyon section (Pierce, 1969; Rice and Langenheim, 1974; Webster, 1969). Also, the Battleship Wash Formation and Indian Springs Formation thin and disappear on the platform. To the northwest of this area, the Mississippian sequence thickens greatly and includes a clastic wedge derived from the Antler Orogenic Belt to the northwest.

Late Carboniferous or Pennsylvanian rocks differ substantially in that redbeds of the Supai Formation of the Grand Canyon are progressively replaced by carbonate rocks in the Callville and Bird Spring Formations of the platform margin and miogeosyncline. The Pennsylvanian rocks also increase in thickness from about 700 feet (210 m) in the North Virgin Mountains (McNair, 1951) to 2,140 feet (650 m) at Arrow Canyon (Cassity and Langenheim, 1966) and to about 2,500 feet (760 m) at Lee Canyon in the Spring Mountains (Rich, 1961).

Subsequent to deposition of the Late Triassic or Early Jurassic Aztec Sandstone and prior to deposition of the early Late Cretaceous Willow Tank Formation in the Muddy Mountains (Longwell et al., 1965), the rocks of the Cordilleran Miogeosyncline were thrust-faulted southeasterly over the platform margin (fig. 28). The most southeasterly of these faults crop out in the Muddy Mountains (figs. 26–28) which are visible from this point. The Muddy Mountain and Keystone Thrusts are the most southeasterly of those in the thrust belt and, presumably, are the sole fault. We will see both of these faults later in the trip and also will note the trace of the Dry Lake and Gass Peak Thrusts to the northwest (fig. 28).

In addition to the thrust faulting, the area of the McCullough Range, River Mountains, Black Mountains, and the southern part of the Virgin Mountains (including the site of this stop) was uplifted subsequent to deposition of the Paleozoic and early Mesozoic platform sequence and before eruption of the Tertiary volcanic rocks around Lake Mead. Thus, the sequence here consists of Tertiary volcanic and sedimentary rocks resting nonconformably on Precambrian crystalline basement (figs. 24–28).

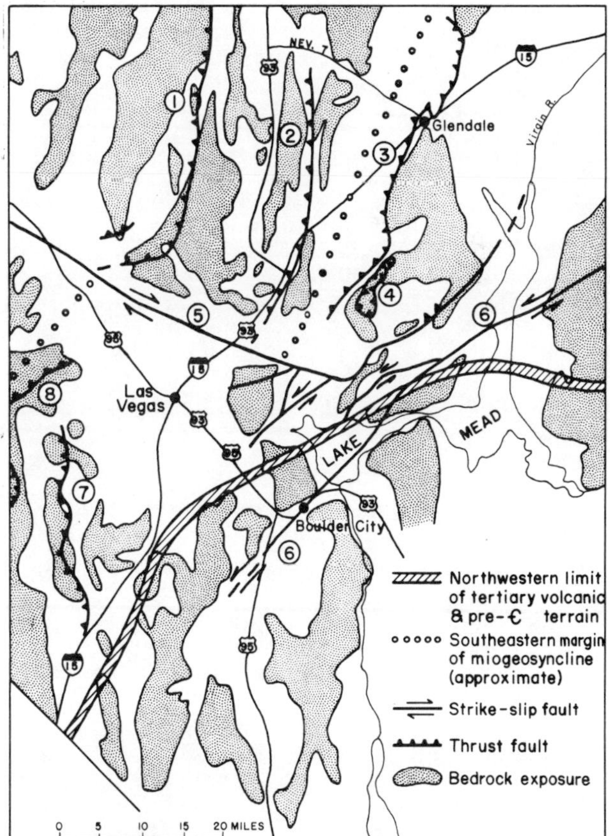

Figure 28. Major structural features and stratigraphic provinces of the Las Vegas-Lake Mead area. Circled numbers on map indicate the following: 1 – Gass Peak Thrust; 2 – Dry Lake Thrust; 3 – Glendale Thrust; 4 – Muddy Peak Thrust and Buffington Pockets; 5 – Las Vegas Valley Shear; 6 – Hamblin Bay – Lime Ridge Shear. Geology compiled after Longwell et al. (1965) and Anderson (1973).

Most of the movement in Basin and Range faulting in this area is younger than the Miocene volcanic rocks of the Lake Mead region and older than the Pliocene Muddy Creek Formation. These faults are not shown on Figure 28, but attention will be called to several of them during the trip. Shear faults of major displacement, however, have been fairly recently discovered (Longwell et al., 1965; Anderson, 1973). The Las Vegas Valley Shear is a right lateral fault inferred to pass along Las Vegas Valley and to separate Frenchman Mountain from the Dry Lake, Las Vegas and Arrow Canyon Ranges (figs. 25, 28). At its southeast end, the trace is lost where it appears to strike into the Black Mountains (figs. 26–28), visible from this point. The sense of movement on the Las Vegas Shear is inferred from deflection of the thrust faults and structures on opposite sides of Las Vegas Valley and from an apparent offset of the boundary between the miogeosynclinal and platform facies as exposed in the bordering ranges (fig. 28). The right lateral, Hamblin Bay–Lime Ridge Fault strikes northeast, passing between Lakeview Viewpoint and the River Mountains (figs. 24, 28) (Anderson, 1973). This fault is also mapped by Anderson (1973) along the northwest slopes of the Black Mountains, onward across the Virgin Arm of Lake Mead and into the Virgin Mountains where it was called the Lime Ridge Fault by Longwell et al. (1965). In addition, presumably it passes through the notch between the Eldorado and River Mountains at Boulder City (fig. 28).

Relationships between the Las Vegas Valley Shear, the Hamblin Bay–Lime Ridge Fault, and the Basin and Range faulting remain somewhat controversial. Longwell et al. (1965) have mapped the Las Vegas Valley Shear more or less as it is shown on Figure 23 and assert that the Frenchman Mountain block probably was moved to its present position from a spot about where the Black Mountains are at present (fig. 23, figs. 26–28). Anderson (1973), however, is of the opinion that the Frenchman Mountain block had shifted to its present position in a southwesterly direction, moving with the northwest block of the Hamblin Bay–Lime Ridge Fault. Transported blocks of rapikivi granite in Tertiary sediments south of the Las Vegas Valley Shear and just east of Frenchman Mountain are the basis for these interpretations, as the only likely bedrock source of such granite is from an intrusion now cropping out in the Virgin Mountains. According to either interpretation, movement on the shear faults is measured in several tens of miles.

ROAD LOG — SIXTH DAY: CLARK COUNTY, NEVADA

Gary D. Webster[1] and Ralph L. Langenheim, Jr.[2]

The road log for the sixth day's excursion in Clark County, Nevada, begins at the intersection of Fremont Street and Las Vegas Boulevard in downtown Las Vegas. We will proceed east to the divide between Frenchman and Sunrise Mountains to examine the uppermost Monte Cristo Group, Indian Springs Formation, and Bird Spring Formation. This section is wholly marine, but relatively thin, and forms part of a so-called "platform sequence" southwest of the Las Vegas Shear. We then will proceed to Webster's (1969) Apex Section, about 9 miles N 10° E of the section at Frenchman Mountain (figs. 22, 23), to examine a much thicker sequence of Indian Springs Formation which contains a notable goniatite occurrence. The Apex section located northeast of the Las Vegas Shear is considered miogeosynclinal. At Apex the Battleship Wash Formation separates the Monte Cristo Group from the Indian Springs Formation and both the Indian Springs and Bird Spring Formations are thicker and more complete. After leaving Apex, we will cross the Las Vegas Valley and climb up out of the desert to Mountain Spring Pass between Potosi Mountain and the main mass of the Spring Mountains. Here we will examine Langenheim's (1956) Mountain Spring Pass section of the Monte Cristo Group.

Total Mileage

0.0 Log begins at intersection of Las Vegas Boulevard and Fremont Street, elev. 2,000 feet (608 m). Proceed northeast on Las Vegas Boulevard.

0.4

0.4 Interstate 15 and U.S. 95 to left, continue straight ahead on U.S. 93 following Las Vegas Boulevard.

1.1

1.5 Stoplight, bear right on U.S. 93 and Las Vegas Boulevard.

0.5

2.0 Stoplight; turn right on Lake Mead Boulevard. Proceed east, past intersections with McDaniel Street at 2.5 miles, Civic Center Drive at 2.8 miles, Peer Road at 3.8 miles, and Lamb Boulvard at 4.8 miles.

3.8

5.8 Stoplight. Intersection with Nellis Boulevard; continue straight ahead. Gass Peak, elev. 6,943 feet, and the south end of the Las Vegas Range are visible to the left, Sunrise Mountain is at 10:30, and Frenchman Mountain, elev. 4,054 feet (1,232 m) is at 12:30.

2.2

8.0 Road turns slightly to left, pull to right and stop alongside road.

STOP 1 — WEST FACE OF FRENCHMAN MOUNTAIN. General view of Cambrian to Mississippian platform.

0.7

8.7 Enter bedrock valley. Precambrian rocks to right, including rapakivi-type granite. Sheared Kaibab-Toroweap (Permian) Formations in Sunrise Mountain block to north (left). Road and pass more or less coincident with a major fault separating Sunrise Mountain and Frenchman Mountain. More or less continuous Cambrian through Early Jurassic section in Frenchman Mountain has figured prominently in both stratigraphic and structural studies of region.

0.3

9.0 Prospect Mountain or Tapeats Sandstone on right is succeeded by green shale of Pioche or Bright Angel Shale. Stratigraphic terminology applied here is derived either from miogeosynclinal sequence in southern and eastern Nevada or from platform sequence in Grand Canyon, as Frenchman Mountain section is geographically and paleogeographically intermediate between these two areas. Prospect Mountain/Tapeats Sandstone is Early Cambrian at this locality, as is lowermost third of Pioche/Bright Angel Shale. Longwell et al. (1965) report between 150 and 400 feet (45–120 m) of basal Cambrian sandstone in Clark County; the lesser thickness is probably more nearly correct for this section. Longwell et al. (1965) report 420 feet (128 m) of Pioche/Bright Angel Shale at this locality. Sheared dolomite of Permian Kaibab and Toroweap Formations is on left or north side of road.

0.1

9.1 Middle Cambrian Lyndon Limestone, about 85 feet thick according to Longwell et al. (1965), forms a hogback separating Pioche/Bright Angel Shale from Chisholm Shale. Open valley to left of road is followed by more sheared dolomite in Kaibab-Toroweap Formations in Sunshine Mountain.

0.5

9.6 Approximate position of Dunderberg "Shale" within post-Chisholm Shale and pre-Devonian carbonate sequence on Frenchman Mountain. Longwell et al. (1965) recognize about 2,200 feet (670 m) of Cambrian carbonate rocks with Middle Cambrian trilobites occurring near base and Late Cambrian trilobites near top. They refer somewhat ambiguously to an occurrence of rocks containing fossils of the Dunderberg Shale, but do not locate them in sequence. Gans (1974) reported approximately 1,050 feet (320 m) of Banded Mountain Member of Bonanza King Formation and about 520 feet (160 m) of Nopah Formation below Sultan Dolomite at this locality, but was unable to identify rocks resembling the Dunderberg Member of the Nopah Formation as he recognized it in the Spring Mountains to the west. Part of the discrepancy between Gans' (1974) report of about

[1] Department of Geology, Washington State University, Pullman, Washington 99163.
[2] Department of Geology, University of Illinois, Urbana, Illinois 61801.

Figure 29. View southward along east slope of Frenchman Mountain. 1 – peak of Monte Cristo Limestone; 2 – dip slopes on top of Bird Spring Formation; 3 – base of Toroweap Formation; 4 – top of cliff-forming member of Toroweap Formation. Kaibab Limestone forms cliff at top of peak behind "4"; bench below cliff is underlain by redbeds belonging to both the Kaibab and Toroweap Formations. 5 – outcropping of Aztec Sandstone, separated by faulting from rocks behind. Strike faulting also somewhat complicates the succession between "2" and "3."

1570 feet (480 m) of rock and the 2,200 feet (670 m) reported by Longwell et al. (1965) is accounted for by the fact that Gans did not describe or measure the rocks of the Papoose Lake Member, the lowermost member of the Bonanza King Formation. The Late Cambrian Nopah Formation is disconformably succeeded by between 500 to 550 feet (150–170 m) (Longwell et al., 1965) of Muddy Peak or Sultan Limestone. McNair (1952), however, reports this unit as approximately 800 feet (245 m) thick at Frenchman Mountain.

0.2

9.8 Cross fault into sheared dolomite of Kaibab-Toroweap Formations in Sunrise Mountain block. Sharply pointed peak to right is composed of Mississippian Monte Cristo Group limestone, which succeeds Muddy Peak or Sultan Limestone in Frenchman Mountain. Monte Cristo Group about 670 feet (205 m) thick at this locality, including about 550 feet (165 m) of Bullion Dolomite, 85 feet (25 m) of Anchor Limestone, and 35 feet (10 m) of Dawn Limestone. Arrowhead Limestone apparently absent and Yellowpine Limestone cannot be distinguished from Bullion Dolomite. At base of Monte Cristo Group, Crystal Pass Limestone, Late Devonian, also apparently absent.

0.2

10.0 Pull off on dirt road and shoulder to right. Double peak at 12 o'clock is composed of Permian Kaibab-Toroweap Formations. To right, south of road, right-hand peak is composed of Monte Cristo Group limestone and left-hand peak of Bird Spring Formation. The valley between is localized by the more easily eroded Indian Springs Formation.

STOP 2 — FRENCHMAN MOUNTAIN. Walk up road to right, entering Indian Springs strike valley on its northeastern side, passing gate and climbing to more or less level spot on Indian Springs Formation. We will discuss the uppermost dolomite of the Monte Cristo Group, the Indian Springs Formation, and much of the Bird Spring Formation (fig. 29), and will present evidence bearing on the Mississippian-Pennsylvanian transition.

0.4

10.4 Pulloff to right is located approximately on line of faulting separating Sunrise Mountain and Frenchman Mountain. Slopes north of road ahead and behind us are composed of Permian Kaibab and Toroweap Formations on Sunrise Mountain block. Outcrops at 10–11 o'clock, south of road, are complexly faulted and their apparent integrity of sequence is misleading. Limestone hogback at 3 o'clock, however, is upper part of Bird Spring Formation and in stratigraphic continuity with rocks observed at Stop 2. Top of Bird Spring Formation is at top of limestone bed which forms low ridge at eastern foot of hogback slope.

0.8

11.2 Old highway; bear right.

0.1

11.3 Turnout on shoulder to right.

STOP 3 — EAST SLOPE FRENCHMAN MOUNTAIN. Turn around and retrace route to intersection of Nellis Boulevard and Lake Mead Boulevard at 5.8 miles on this log. In passing through pass between Frenchman Mountain and Sunrise Mountain, higher level alluvial fans are well displayed as viewed going westerly.

5.4

16.7 Intersection of Nellis Boulevard and Las Vegas Boulevard; stoplight. Turn right to travel north on Nellis Boulevard.

2.7

19.4 Intersection of Nellis Boulevard and Las Vegas Boulevard N. Turn right and proceed northeast.

0.8

20.2 Traffic light at main entrance to Nellis Air Force Base. Continue straight ahead.

0.4

20.6 Railroad crossing and auxiliary entrance to Nellis Air Force Base.

2.5

23.1 Entrance to Lake Mead Base, U.S. Navy. Base buildings are visible to right of northern foot of Sunrise Mountain.

0.1

23.2 Railroad crossing, Lake Mead Base spur of Union Pacific Railroad.

2.8

26.0 Dune tract at 3 o'clock, southeast of road. Terraced slopes to north in Las Vegas and Arrow Canyon Ranges are almost entirely composed of Bird Spring Formation and are characteristic of outcrops of this formation in the desert environment.

1.6

27.6 Overpass carrying I–15; continue straight ahead on old U.S. 93.

0.2

27.8 Underpass beneath main line, Union Pacific Railroad.

0.5

28.3 Monte Cristo Group exposed in roadcut.

0.2

28.5 Turn left on gravel road. Bird Spring Formation forms terraced outcrops on hill at skyline at 11 o'clock.

0.2

28.7 Turn left, leave pole-line road to take a graded road giving access to an airway beacon.

0.5

29.2 Turn left on desert track, several anastamosing trackways proceed generally westward into wash at this point.

0.1

29.3 Desert track continues over edge of scarp bounding wash to west; continue with caution and with consideration of off-road capability of your vehicle. This point is 0.3 miles from the "end of the line" on this spur of the road log. It may be worth it to walk! Drainage line descends toward this point from right in a wash separating Bird Spring Formation outcrops from alluvium. Wash turns to right, away from this point, to receive a tributary from west. Tributary wash flows in a strike valley in Indian Spring Formation and Bird Spring Formation, with dip slopes of Battleship Wash Formation and Yellowpine Limestone forming south wall of valley. Our traverse proceeds up this strike valley to first tributary on north side. Vehicles with low clearance should stop here.

0.1

29.4 Enter wash at track crossing. Proceed across wash and up strike valley on its northern bank.

0.2

29.6 Prospect tunnel to right of track. This is a good place to turn around and is the end of our logged route to this stop.

STOP 4 — APEX SECTION. Yellowpine Limestone, Battleship Wash Formation, Indian Springs Formation and basal portion of Bird Spring Formation are accessible at this locality. Principal purpose of stop is to visit a notable goniatite locality in Indian Springs Formation, which may be reached by proceeding up wash on foot to first branch. At this point, bear right and continue to next major fork. At that point bear left and continue up wash to a low cliff formed by a vertical bed of fine-grained quartzite. Goniatites are in a calcareous lens at top of conglomerate just above quartzite. Return to your vehicles and retrace log to I–15.

1.8

31.4 Underpass beneath Union Pacific Railroad.

0.2

31.6 Exit right to enter I–15.

0.2

31.8 Enter I–15, traveling west.

5.0

36.8 Las Vegas Range and Gass Peak at 3 o'clock, Sunrise Mountain at 9 o'clock with U.S. Navy Lake Mead Base at foot of mountain to left. This base is for storage of nuclear devices. Mountains on horizon in far distance are Black Mountains on Arizona side of Lake Mead. Potosi Mountain, elev. 8,504 feet (2,585 m), in southern end of Spring Mountains, is a prominent peak at 12 o'clock.

2.4

39.2 Road curves to left. Higher level alluvial fan surface, capped by caliche, is preserved at top of remnant hills along road.

3.1

42.3 Alluvial fans to right are especially well developed. Smaller fan at 9:30 heads against Gass Peak, highest point on horizon. Webster's (1969) Gunnery Range section is to right of this fan and his Gass Peak Section is to left. An especially well-developed fan heads against south end of Las Vegas Range at 4 o'clock.

4.6

46.9 Exit for U.S. 95 South. Access to Hoover Dam. Continue straight ahead on I–15. Take middle lane and continue southwesterly, past strip, to Blue Diamond-Pahrump exit 33. Note slopes of Spring Mountains and La Madre Mountain to right (west), beyond Las Vegas Valley. Keystone Thrust, southeasternmost fault in thrust belt at this point, crops out along these slopes from Potosi Mountain to La Madre Mountain. A more or less continuous sequence, though faulted, of Kaibab-Toroweap Formations, Moenkopi Formation, Chinle Formation, and Aztec Sandstone crops out below overthrust. Aztec Sandstone forms prominent red and pink cliffy slope low on mountain face. Goodsprings Dolomite, Sultan Limestone, Monte Cristo Group, and Bird Spring Formation, in upper plate, form tree-covered slopes above Chinle-Aztec cliffs. Goodsprings Formation (Group) in Spring Mountains includes Bonanza King Formation, Nopah Formation, Mountain Springs Formation and Pogonip Formation according to Gans (1974) and ranges from Late Cambrian through Early Ordovician in age.

8.8

55.7 Exit 33, Blue Diamond-Pahrump; leave freeway to right.

0.5

56.2 Turn right on Blue Diamond-Pahrump Road, heading west.

0.2

56.4 Enter Blue Diamond Road heading west. Bird Spring Range, type locality of Bird Spring Formation is at 10–11 o'clock. Again note characteristic terracing of Bird Spring Formation outcrops. Potosi Mountain is at 1 o'clock. Ridge of Kaibab-Toroweap limestone at 2 o'clock with Aztec Sandstone cliffs beyond.

2.6

59.0 Union Pacific Railroad, mainline of Los Angeles and Salt Lake Railroad. This line, constructed in 1905 and financed by Jay Gould, was the last major element of our primary rail system to be constructed.

1.7

60.7 Railroad crossing, Union Pacific Blue Diamond spur.

0.4

61.1 Pulloff to right. Kaibab-Toroweap Formations crop out on left side of road. Bird Spring Range at 11 o'clock with Potosi Mountain beyond. Good view of Aztec Sandstone cliffs below Keystone Thrust at 12 o'clock. White scars on ridge at 1 o'clock are gypsum mines. Base of La Madre Mountain visible at 2:30.

0.7

61.8 Permian Kaibab-Toroweap Formations to right of road, succeeded by Triassic Moenkopi Formation.

2.9

64.7 Moenkopi redbeds with included limestone units to right of road in low hills.

1.7

66.4 Pulloff to right. Low rounded hills in front of and below Kaibab-Toroweap cliffs at 12–1 o'clock are outcrops of olive-green, thinly bedded limestone within Triassic Moenkopi Formation. These rocks contain *Pentacrinus* columnals and ammonites. They are separated from Permian rocks apparently above them by a thrust fault trending parallel to strike.

0.3

66.7 Blue Diamond Road diverges to right. Continue, bearing left, on Pahrump Road toward Mountain Springs Pass.

1.1

67.8 Low rounded hill at 3 o'clock is of limestone in Triassic Moenkopi Formation. It is separated from cliff of Permian limestone to west by a thrust fault.

0.3

68.1 Supai Group redbeds below limestone of Kaibab and Toroweap Formations at 2 o'clock.

0.3

68.4 Moenkopi Formation limestone above thrust fault is visible through a gap in Kaibab-Toroweap ridge at 1 o'clock.

0.4

68.8 Permian limestone of Kaibab and Toroweap Formation crops out on both sides of road.

0.3

69.1 Faulted Triassic and Permian rocks to right. We are entering an area with abundant Joshua trees (*Yucca brevifolia*). *Cholla*, a shrub-like, abundantly spinose cactus, occurs with Joshua trees, along with *Yucca mohavensis*, a yucca having longer leaves, but lacking tree-like form of the Joshua tree. The small shrub with yellow-green leaves is creosote bush (*Larrea tridentata*), which gives a characteristic odor when its leaves are crushed. The large shrub in the washes is mesquite (*Prosopis chilensis* and/or *P. pubescens*), a very spiny legume which provided food for the Indians formerly inhabiting this region. Mesquite is something of a nuisance at the Arrow Canyon locality to be visited tomorrow. *Larrea* is the dominant desert shrub in the lower, hotter regions of the desert, as you have seen in the regions so far traversed. The Joshua tree is characteristic of the upper, cooler desert and of the transition into the so-called pygmy forest of *Pinus edulis* (nut pine) and *Juniperus utahensis* (juniper) which covers the lower slopes of the Spring Mountains ahead of us. *Yucca*, incidentally, is a lily.

2.6

71.7 Aztec Sandstone forms pink cliffs ahead and rests on bright-red Triassic Chinle Formation. These rocks are in lower plate of Keystone Thrust.

1.0

72.7 Pulloff to right

STOP 5 — VIEW OF KEYSTONE THRUST.

1.6

74.3 Cross fault line separating Aztec Sandstone from Paleozoic carbonate rocks. Road very shortly enters Devonian terrain.

0.7

75.0 Pulloffs to left and right. Thinly bedded, very fine-grained limestone in roadcut is Crystal Pass Limestone, entirely or almost entirely of Late Devonian age. Dark-colored, ledgy slope above is formed in Dawn Limestone and lighter colored, cherty cliff above slope is Anchor Limestone. This is a good place to examine Devonian-Mississippian contact and to collect fossils in Dawn and Anchor Formations of Monte Cristo Group. Crystal Pass Limestone is a very fine-grained, pelletoidal limestone. It is very poorly fossiliferous, but, elsewhere, has yielded a sparse fauna of conodonts and foraminifers. Overlying Dawn Limestone Member of Monte Cristo Group is composed of grain supported, pelletoidal biocalcarenite with abundant fossils. Contact between two units is abrupt and, at microscopic level, appears to be a disconformity in Arrow Canyon Range, where it has been examined in detail. In Arrow Canyon Range, Famennian conodonts are reported in upper part of Crystal Pass Limestone (Langenheim and Collinson, 1963; Reinbold and Langenheim, 1974). Brenckle (1973), however, reports late Kinderhookian endothyroid foraminifers from uppermost Crystal Pass Limestone in essentially the same locality. Relevant localities in Arrow Canyon Range will be seen at a distance, from highway, near the end of tomorrow's excursion under conditions precluding a visit to the outcrops.

0.1

75.1 Cross normal fault to enter outcrop of Goodsprings Dolomite — the thick-bedded, rubbly weathering material. Hewett (1931), in his reconnaissance report, did not subdivide these rocks or correlate them with described formations known elsewhere in the Great Basin. Gans (1974), however, has now subdivided the Goodsprings Dolomite and recognizes approximately 1,090 feet (330 m) of Banded Mountain Member of Bonanza King Formation; 660 feet (200 m) of Nopah Formation; and about 340 feet (100 m) of Mountain Springs Formation within Goodsprings Dolomite mapped by Hewett at Mountain Springs Pass. Bonanza King Formation and Nopah Formation are Late Cambrian, Mountain Springs Formation is a local equivalent of Middle and Late Ordovician Ely Springs Dolomite. Basal Papoose Lake Member of Bonanza King Formation is absent at base of sequence along Keystone Thrust according to Gans (1974). Eureka Quartzite and Pogonip Group also are absent here because of pre-Mountain Springs Formation erosion. This sequence is a platform section, which, in many respects, resembles the succession at Frenchman Mountain.

Here we are well within the Pinyon-Juniper or "Pygmy Forest," the ecological zone just above the desert. The "cedars" are *Juniperus utahensis* and the pines are the nut pine, *Pinus edulis*, which yields pine nuts (piñones), a major element in the indigenous Indian diet and a delicacy of contemporary Mexican cookery. Shrubs include true sagebrush, *Artemisia tridentata*, scrub oak, *Quercus dumosa*, and *Cercocarpos*, none of which is found in the true desert below about 5,000 feet (1520 m) elevations.

1.5

76.6 Crest of Mountain Springs Pass, elev. 5,350+ feet (1626 m). Rounded hill at 12:30 is site of Langenheim's (1956) Mountain Springs Pass section (fig. 30). Crest of hill is in Bullion Formation of Monte Cristo Group, here a dolomite. Dark rock on bench just this side of summit is outcrop of Dawn and Anchor Limestones of Monte Cristo Group. Light-gray–weathering rock below is in Crystal Pass Limestone which here rests on Valentine Member of Sultan Limestone in a gradational contact. Prominent brown-weathering ledge on brow of slope is a 6-foot (2 m) layer of sandstone about 85 feet (25 m) below top of Valentine Member. Lower slopes of hill are occupied by remainder of Valentine Member and Ironside Member of Sultan Formation according to Hewett (1931). Mississippian Arrowhead Limestone, Yellowpine Limestone, and Indian Springs Formations are exposed along crest of ridge to north and out of sight at this point. Fossil specimens collected from this site by the Wheeler expedition included brachiopods identified and illustrated by C. A. White (1874) as *Spirifer centronatus*. This particular fossil subsequently became an index for sub-Carboniferous (i.e. Mississippian) rocks throughout the western states.

West slope of Potosi Mountain well displayed at 7 to 8 o'clock. Sultan Limestone forms lower slopes but peak and upper part consist entirely of Monte Cristo Group involved in repeated thrust faulting. Three superposed cliffs of Bullion Dolomite may be distinquished from this point.

0.5

77.1 Turnoff for Mountain Springs Lodge and base of section.

STOP 6 — MOUNTAIN SPRINGS PASS. Park here or at turnoffs farther west along road and climb ridge north of road to examine Monte Cristo Group. Buff-weathering layer just above black cliff at south end of ridge is sandstone layer within Valentine Member of Sultan Formation. Base of Dawn Limestone is at false summit as seen from entrance to Mountain Springs Lodge. At 12 o'clock Kaibab and Toroweap limestones form cliffs on skyline above red slopes of Supai Group. Low hill displaying characteristic terraced weathering form of Bird Spring Formation is visible low in notch of stream along road.

0.2

77.3 Prominent cliff, halfway up slope at 2 o'clock, is basal Yellowpine Limestone with a small outcrop of Arrowhead Limestone beneath cliff. This is the only outcrop of Arrowhead Limestone in the area. Crag to left of cliff on skyline is a bed of limestone near base of Indian Springs Formation. At 3 o'clock a drainage line more or less follows Dawn-Anchor contact. Dark rock is an outcropping of these two formations, the cherty, upper material being Anchor Limestone. Crestline is in dolomitized Bullion Formation.

0.3

77.6 Indian Springs Formation-Yellowpine Limestone contact at 3 o'clock.

0.2

Figure 30. Mountains Springs Pass stratigraphic section as seen from crest of Mountain Springs Pass. 1 – cliff-forming ledge of black dolomite in Ironside Member of Sultan Formation. Light-colored rock below is in top of Goodsprings Dolomite. 2 – sandstone layer in Valentine Member of Sultan Formation; 3 – transition from Crystal Pass Limestone into Dawn Limestone of Monte Cristo Group; 4 – approximate contact between Dawn and Anchor Limestones; 5 – approximate base of Yellowpine Limestone. Crest of hill is in Bullion Formation.

Figure 31. View of upper portion of Mountain Springs Pass stratigraphic section from end of road log. 1 – base of Bullion Formation of Monte Cristo Group; 2 – about at base of Yellowpine limestone; 3 – near contact with Yellowpine Limestone of Monte Cristo Group and Indian Springs Formation; 4 – approximately at contact between Indian Springs Formation and Bird Spring Formation.

77.8 Poorly to reasonably well-exposed Bird Spring Formation at 11:00 to 4:30.

0.8

78.6 Enter strike valley formed in upper part of Bird Spring Formation and in Supai Group. Kaibab-Toroweap resting on top of sequence at 12 o'clock. Note steeply dipping Bird Spring Formation at 9 o'clock, upper part of Mountain Spring Pass Section at 12 o'clock (fig. 31).

Retrace route crossing I–15, turning north toward Las Vegas on old highway 91 which leads directly into south end of Strip.

END OF SIXTH DAY ROAD LOG

STOP DESCRIPTIONS — SIXTH DAY

R. L. Langenheim, Jr.[1] and Gary D. Webster[2]

STOP 1 — WEST FACE OF FRENCHMAN MOUNTAIN
NW¼, sec. 23, T. 20 S., R. 62 E.
Clark County, Nevada

Precambrian through Late Mississippian and/or Early Pennsylvanian rocks are exposed in the west face of Frenchman Mountain ahead of us and south of the road (fig. 32). Bedrock at the head of the wash that leads from the canyon responsible for the first prominent notch south of the road on the skyline of Frenchman Mountain is Precambrian crystalline rock. The water tank at the foot of the slope to the right at 2:30, taken facing up the road, is at the head of an alluvial fan abutting Cambrian rocks in the bedrock slope above. The large white letter "C" just below the white-weathering ledge is painted on the Early to Middle Cambrian Pioche Shale or the lower part of the Bright Angel Shale. The lowermost third of the Pioche Shale is Early Cambrian in age. The bed below the shale is Early Cambrian Prospect Mountain or Tapeats Sandstone and the white layer above the "C" is probably the Middle Cambrian Lyndon Limestone. The bench above the Lyndon Limestone belongs to the Chisholm Shale or the upper part of the Bright Angel Shale. The lowermost 2,200 feet (670 m) of limestone and dolomite above the Chisholm Shale are Middle and Late Cambrian equivalents of the Bonanza King Formation (Longwell et al., 1965). They are succeeded by a Dunderberg Shale equivalent and a thin dolomite layer, probably part of the Late Cambrian Nopah Formation (Longwell et al., 1965). The Late Devonian sequence above, variously referred to the Muddy Peak Limestone or Sultan Limestone, is 500 to 550 feet (150–170 m) thick (Longwell et al., 1965) and rests disconformably on the Late Cambrian Nopah Formation. The Mississippian Monte Cristo Group, 670 feet (205 m) thick, rests disconformably on the Devonian. It includes the Dawn Limestone, Anchor Limestone, and undifferentiated Bullion Dolomite and Yellowpine Dolomite. The top of the second peak south of the road (the one with aerials) is in the lowermost, Late Mississippian and Early Pennsylvanian part of the Bird Spring Formation. The Bird Spring Formation rests on a thin sequence of the Late Mississippian Indian Springs Formation, which forms a narrow bench above the thick, massive cliffs of the combined Bullion and Yellowpine Dolomites.

Provided atmospheric conditions are good, which is unlikely after midmorning, good views of the McCullough Mountains, the Bird Spring Range, Potosi Mountain, La Madre Mountain, Charleston Peak, Las Vegas Valley, the Sheep Range, and Gass Peak (fig. 23) and of their general geology and major structural relations (fig. 29) may be had from this point. At 9 o'clock relative to a stance facing down the road, the sharp peak in the McCullough Range is of volcanic rock. This entire range is in the terrain of Tertiary volcanic rock and sedimentary rock resting on Precambrian basement, characterizing the platform southeast of this point (fig. 28) which includes all of the terrain traversed yesterday between Kingman, Arizona, and Railroad Pass at mile 13.5 on the Hoover Dam–Las Vegas road log.

Potosi Mountain is the high peak at 10:30, just behind the strip. The crest of Potosi Mountain is composed of Devonian Sultan Limestone, but the remainder of the mountain is composed of intricately faulted Mississippian Monte Cristo Limestone and other Paleozoic rocks of the platform margin facies. We will pass along the northern foot of Potosi Mountain later today between Stops 5 and 6 (figs. 22, 23). Potosi Mountain, incidentally, was the site of an airline crash in which Carole Lombard was killed in 1941. The low, rounded, brown-gray hills to the left of Potosi Mountain include the Bird Spring Range. This is the type locality of the Bird Spring Formation of Hewett (1931). At the south end of these hills, Mississippian Monte Cristo Group and Devonian Sultan Limestone crop out, including the Dawn Limestone being quarried at the Sloan Quarry for limestone and dolomite. The white scar on the skyline at about 10 o'clock is part of the quarry. The gap to the south, separating the McCullough Range from the Bird Spring Range, is occupied by the Union Pacific Railroad and I–15 en route to California and Los Angeles. It also is the boundary between the Tertiary volcanic and Precambrian terrain to the southeast and a terrain of platform margin Paleozoic and early Mesozoic rocks to the northwest (fig. 28).

The highest peak on the skyline at 1 o'clock is Charleston Peak (elev. 11,912 ft (3,621 m)) which is the highest peak in the Spring Mountains and is by far the highest point in southern Nevada. The peak and the ridgeline to the south are composed of Bird Spring Formation and the range itself is the site of Longwell and Dunbar's (1936) reference section for the Bird Spring Formation. Longwell and Dunbar (1936) named the Indian Springs Formation as a member of the Bird Spring Formation and first recognized the presence of Late Mississippian through Leonardian rocks within the Bird Spring Formation.

Prominent pink and red cliffy outcrops compose much of the mountain face halfway between Charleston Peak and Potosi Mountain. These are Triassic Chinle and Late Triassic or Early Jurassic Aztec Sandstone and are exposed just below the Keystone Thrust Fault. The Keystone Thrust (figs. 22, 28) is the sole fault of the thrust belt bordering the southern Nevada part of the Cordilleran Miogeosyncline. We will view this fault at Stop 5, later today. The low, dull brown-gray ridge in front and to the left of the redbed outcrops is formed of a Permian Supai through Kaibab Formation sequence stratigraphically below the Triassic Moenkopi Formation, which is largely hidden from view. The dark, tree-covered slopes above the Aztec Sandstone are underlain by Goodsprings Dolomite through Bird Spring rocks that are indistinguishable from this distance. These rocks are in the overthrust block of the Keystone Thrust.

The prominently banded outcrops directly in front of Charleston Peak are chiefly composed of Mississippian through Permian Bird Spring Formation, but the strike ridge trending directly toward us at 12:30 includes a nearly complete Paleozoic sequence on the upper plate of the Keystone Thrust. This is La Madre Mountain (fig. 22). The very prominent, light-gray outcrop band, inclined at an apparent dip of about 30° N and

[1] Department of Geology, University of Illinois, Urbana, Illinois 61801.
[2] Department of Geology, Washington State University, Pullman, Washington 99163.

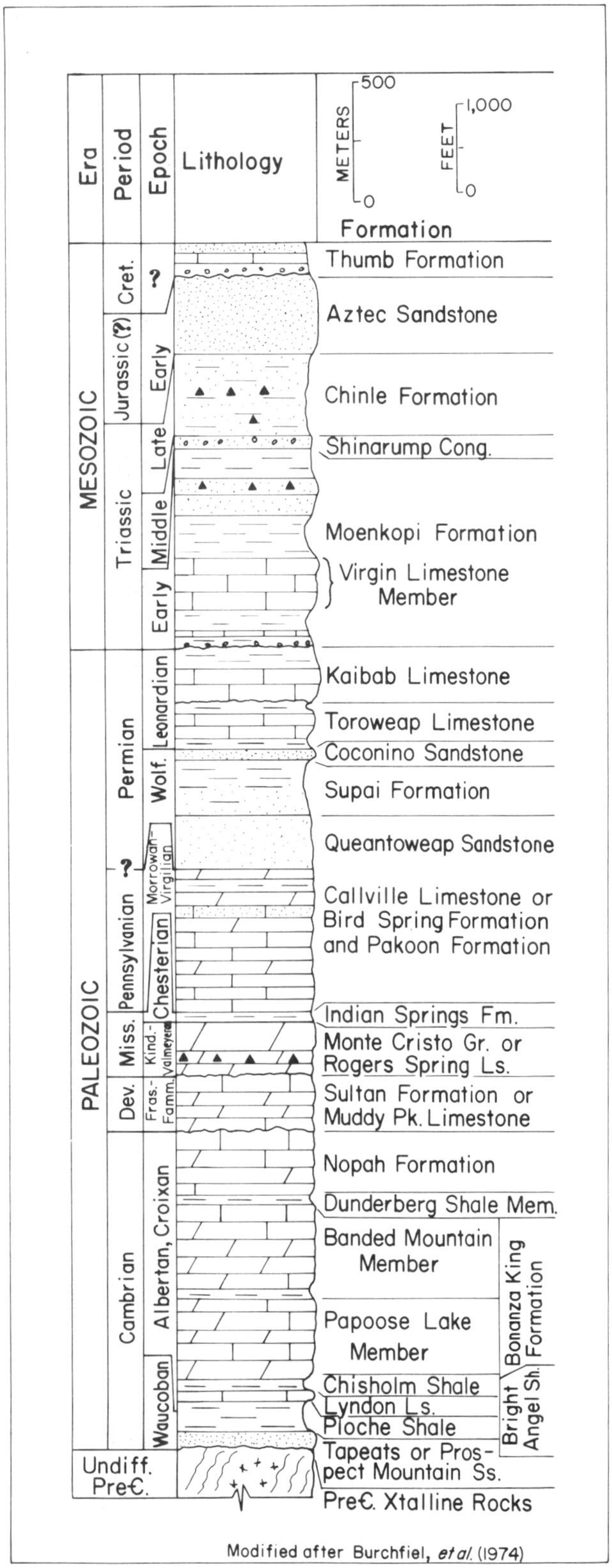

Figure 32. Columnar section of Cambrian through Jurassic platform sequence exposed on Frenchman Mountain. Unit thicknesses shown are after Figure 2 of Burchfiel et al. (1974); formational terminology encompasses all names widely applied to these units at this place. Terminology and thickness figures are not standardized because this sequence has yet to be studied in detail and in its entirety.

exposed on the right side of the ridge at its contact with alluvium, is an exposure of the Bullion Formation of the Monte Cristo Group. The valley north of this ridge is Kyle Canyon and the next valley northward is Lucky Strike Canyon. Lucky Strike Canyon is inconspicuous in this view, but the next conspicuous canyon is Lee Canyon. Rich's (1961) stratigraphic section of the Bird Spring Formation was measured in the hills north of Lee Canyon which form the skyline from this viewpoint.

The Las Vegas Valley lies northeast of the Spring Mountains, separating them from the Sheep Mountains and Gass Peak. The valley extends to the northwest at about 1:30 where alluvium forms the skyline. This valley contains the trace of Longwell's (1960) Las Vegas Shear (figs. 22, 28), the presence of which is inferred entirely from apparent right-lateral displacement of major structures, facies patterns, and stratigraphic distribution on opposite sides of the valley.

The highest peak northeast of Las Vegas Valley, at about 2:30, is Gass Peak (elev. 6,943 ft (2,111 m)), which is at the south end of the Las Vegas Range. The mountain consists almost entirely of Monte Cristo Group and Bird Spring Formation rocks, with the dark material at the base being about 2,000 feet (608 m) of miogeosynclinal Middle to Late Devonian rocks. The dark, tree-covered skyline beyond Gass Peak and to the right is the crest of the Sheep Range (fig. 22). Here the rocks mostly belong to a miogeosynclinal sequence of more than 3,000 feet (910 m) of Early through Late Ordovician rocks. A Cambrian sequence more than 7,000 feet (2,100 m) thick is largely hidden from this point, as are the more than 1,000 feet (300 m) of Silurian rock. These thicker and more complete miogeosynclinal rocks will be viewed in large part at Stops 5 and 6 tomorrow, and are in marked contrast to the section in Frenchman Mountain and at Mountain Springs Pass, which we see today.

Banded outcrops of Bird Spring Formation continue at the head of a large alluvial fan at 3 o'clock. The dark rounded hill to the left of the fan is composed of Devonian rocks and Monte Cristo Group. Webster's (1969) Gass Peak section is just northwest of this hill and behind it. His Gunnery Range Section is to the right of the fan. All of the banded outcrops to the southeast are in Bird Spring Formation and the apparent anticline bordering a shallow syncline to its southeast is in the south end of the Arrow Canyon Range (fig.22).

The checkered watertowers in the foreground to our north are on the Nellis Air Force Base, a principal advanced training base for tactical aircraft. The activities of these aircraft will probably command your attention from time to time during our stay in Clark County.

The Gass Peak Thrust (figs. 22, 28) separates Gass Peak from the Sheep Range. It is the second of two major well-documented thrusts cropping out northwestward of the Muddy Peak Thrust in the block northeast of the Las Vegas Shear. The Muddy Peak Thrust is the sole fault in the thrust belt north of the Las Vegas Shear and is a presumed equivalent to the Keystone Thrust, thus suggesting that movement on the Las Vegas Shear is at least as great as the distance between outcrops of the Muddy Peak Thrust and the Gass Peak Thrust.

Frenchman Mountain and Sunrise Mountain comprise an isolated block of Paleozoic and early Mesozoic platform-sequence rocks closely adjacent to the presumed trend of the Las Vegas Shear to the northeast. The west side of the block is presumably bounded by high-angle, normal faults of the Basin and Range type and the interior of the block is intricately subdivided by faults of this type (Longwell et al., 1965; Longwell, 1960; Anderson, 1973). The south end of the range also is cut off by a presumed right-lateral fault that strikes subparallel to the Las Vegas Shear (Longwell, 1960; Longwell et al., 1965; Anderson, 1973). The east side of the Frenchman

Mountain-Sunrise Mountain block, however, is shown as a dip slope by Longwell (1960) and Longwell et al. (1965), but is presumably cut off by a left-lateral fault branching from the Las Vegas Shear according to Anderson (1973). Figures 22 and 28 reflect these conflicting interpretations. Both the structural geology and the stratigraphy of the Frenchman Mountain-Sunrise Mountain block are pivotal to understanding the regional geological relationships.

STOP 2 — FRENCHMAN MOUNTAIN
SE¼, sec. 24, T. 20 S., R. 62 E.
Clark County, Nevada

Carboniferous strata form a double peak with an intervening strike valley and saddle on the north peak of Frenchman Mountain (fig. 33). Three stratigraphic units are recognized: Monte Cristo Group (west peak), Indian Springs Formation (strike valley and saddle), and Bird Spring Formation or Callville Limestone of Longwell et al., 1965 (east peak). Strata strike slightly west of north and dip steeply eastward, approximately 45-50°. The complete section is considered to be platform facies and is probably displaced approximately 20 to 25 miles (32-40 km) west of its Carboniferous paleogeographical position (Longwell, 1960).

The Monte Cristo Group at Frenchman Mountain (fig. 34) is only 672 feet (205 m) thick in contrast to 832.5 feet (253 m) at Mountain Springs Pass and 1,099 feet (334 m) at Arrolime. In addition, here the entire Monte Cristo Group is dolomitized and the Arrowhead Limestone has not been distinguished. In thickness the sequence more closely resembles the Redwall Limestone as described by McKee and Gutshick (1969) in the Grand Canyon. In addition, the Dawn-Anchor-Bullion Formation sequence is reminiscent of the Whitmore Wash-Thunder Springs-Mooney Falls Member succession. Dolomite, however, becomes less prominent in the western part of the Grand Canyon as the Frenchman Mountain locality is approached. The Bullion Formation is dolomitic at Mountain Springs Pass, to the west, and Arrolime, to the northeast, and the overlying Arrowhead Limestone and Yellowpine Limestone are dolomitic at Arrolime. Farther to the north in the Arrow Canyon, Las Vegas, and Meadow Valley Ranges, and to the northeast in the Mormon, Virgin, and Muddy Mountains, the entire Monte Cristo Group is composed of limestone (Langenheim et al., 1962; Langenheim, 1963; Pierce and Langenheim, 1972). The pattern of dolomitization suggests that it either is related to the extreme shelf margin or, possibly, to proximity to the thrusts and shear faults of the area.

Depiction of the Indian Springs and Bird Spring Formations in Figure 33 is generalized as a result of relatively poor exposures at the site of the measured section and disinterest in the post-Monte Cristo Group sequence at the time of measurement.

The Indian Springs Formation (fig. 35) is 85 feet thick near the north end of Frenchman Mountain and consists of thin-bedded sandstone, siltstone, limestone, dolomite and conglomerate. Conodonts indicate that the Mississippian-Pennsylvanian boundary is within a conglomerate between 54 and 60 feet (16–18 m) above the base. In the base of the conglomerate, conodonts of the *Rhachistognathus muricatus* Zone include *R. muricatus, Gnathodus bilineatus, G. commutatus commutatus, G. girtyi simplex, Cavusgnathus naviculus,* and *C. convexus,* all of which are considered Chesterian (Webster, 1969). Only *R. muricatus* ranges into the Morrowan. The upper part of the conglomerate yields a few specimens of new species of *Idiognathoides* and *Idiognathus*?. These are the oldest recognized species of these two genera in the Great

Figure 33. The Mississippian-Pennsylvanian transition at Stop 2. View southwest from terrace on roadway at its crossing of Indian Springs Formation. 1 – base of Indian Springs Formation; 2 – top of Indian Springs Formation; 3 – base of "Tungsten Gap Chert"; 4 – top of "Tungsten Gap Chert."

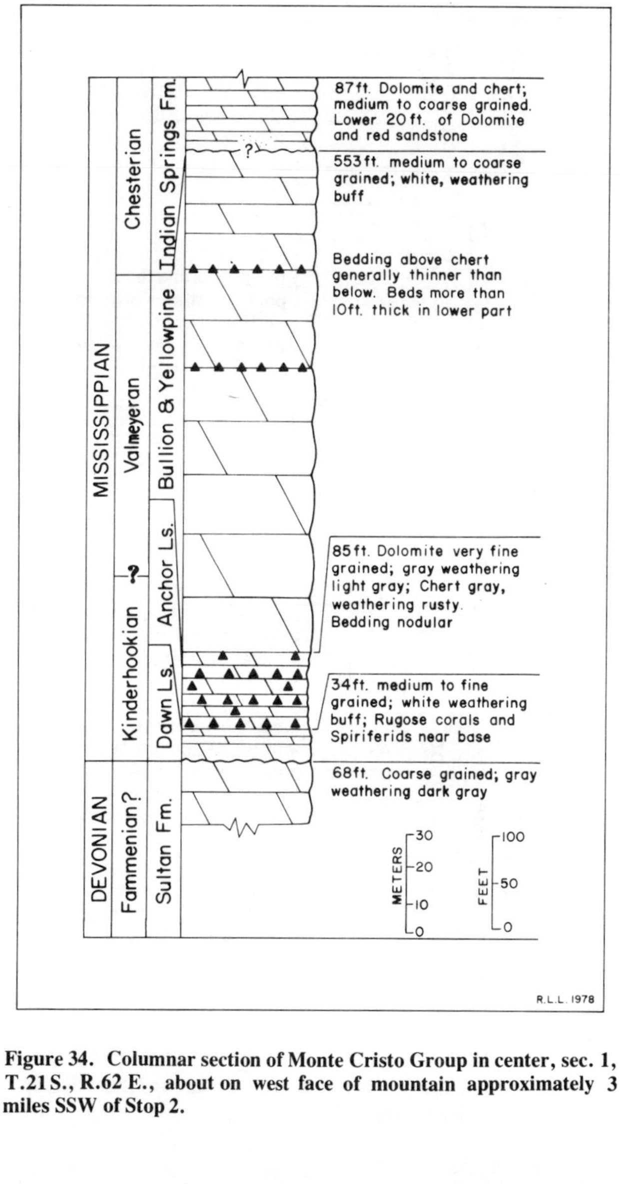

Figure 34. Columnar section of Monte Cristo Group in center, sec. 1, T.21 S., R.62 E., about on west face of mountain approximately 3 miles SSW of Stop 2.

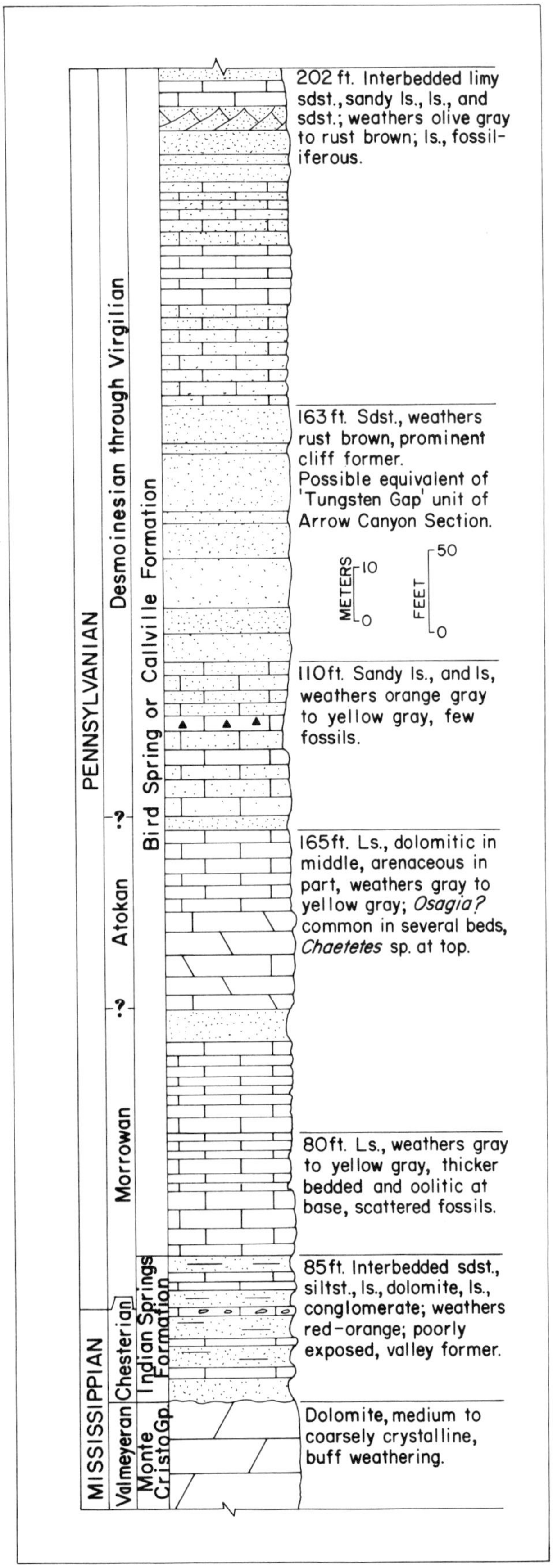

Figure 35. Columnar section of Indian Springs and Bird Spring or Callville Formation at Stop 2.

Basin. Limestone at 65 feet (20 m) above the base (5 feet (1.5 m) above the conglomerate) yields *Rhachistognathus primus* and *R. muricatus* among other conodonts. The *R. primus* Zone is considered basal Pennsylvanian (Lane and Straka, 1974). Both right and left forms of *Adetognathus lautus* occur throughout the conglomerate and in the overlying limestone. Lane, Sanderson and Verville (1972) and Lane and Straka (1974) consider *A. lautus* indicative of basal Pennsylvanian in the southern Midcontinent. Furthermore, Lane, Sanderson and Verville (1972, p. 552) report *A. lautus* as occurring above the *Rhipidomella nevadensis* Zone (brachiopod) in southern Nevada. At Apex, Nevada (Stop 4, today's log) and Arrow Canyon (Stop 3, tomorrow's log) as well as several other localities in southern Nevada (Webster, 1969) and in the Oquirrh Mountains of Utah (Webster, unpublished data) *A. lautus* occurs within the uppermost part of the *R. nevadensis* and *Rhachistognathus muricatus* Zones. This suggests that *A. lautus* may have evolved in the marginal basins and then spread onto the Midcontinental stable platform during Early Pennsylvanian trangressions.

West and north of Frenchman Mountain, in the basinal facies, the Mississippian-Pennsylvanian boundary is recognized within the lower part of the Bird Spring Formation (Webster and Lane, 1967; Webster, 1969; Dunn, 1970; Lane, Sanderson and Verville, 1972; Lane and Straka, 1974). Thus, the Indian Springs Formation appears to be time transgressive to the east.

The Callville Limestone consists of 720 feet (220 m) of limestone and sandstone with the lower 355 feet (108 m) predominantly carbonate and the upper 365 feet (110 m) predominantly sandstone. The prominent brown-weathering cliff-forming sandstone in the basal part of the upper 365 feet (110 m) is probably equivalent, at least in part, to the "Tungsten Gap Chert" of Castle (1967) and of Micklin (1967) to be viewed at Arrow Canyon tomorrow. Longwell et al. (1965) recognized about 1,500 feet (455 m) of Callville Limestone at Frenchman Mountain. This includes the Indian Springs Formation and Bird Spring Formations as here recognized and the Permian Pakoon Formation as recognized by McNair (1951).

Although series boundaries are not firmly established in the Callville Limestone, conodonts suggest that the Morrowan-Atokan boundary is in or near a sandy interval 140 feet (42 m) above the base. This is supported by the occurrence at 130 feet (40 m) and 145 feet (44 m) above the base of plates of *Paragassizocrinus calyculoides*, a late Morrowan and early Atokan crinoid widespread in the eastern Great Basin, and *Anchignathus coloradoensis* at 161 feet (49 m) above the base. A *Chaetetes* biostrome 245 feet (75 m) above the base is probably near the Atokan-Desmoinesian boundary. The Desmoinesian-Missourian and Missourian-Virgilian boundaries are not established; however, Brill (1963, fig. 9) suggested that the Desmoinesian is absent and reported *Triticites neglectus* about 525 feet (160 m) above the base of his section. To the east there is a widespread unconformity at the base of the Virgilian, and Missourian strata are lacking or were removed by erosion (McKee, 1975). Conodonts similar to Midcontinent forms, including *Streptognathus elegantulus, Adetognathus* n. sp., *Neognathodus* n. sp. and *Anchignathodus* cf. *A. edentulus*, occur in the upper 115 feet (35 m) of the Callville and suggest a Virgilian age.

STOP 3 — EAST SLOPE FRENCHMAN MOUNTAIN
Unsurveyed, approximately NE¼, NE¼, sec. 19, T. 20 S., R. 63 E.
Clark County, Nevada

Red bouldery outcrops at 12 o'clock are Late Triassic or

Early Jurassic Aztec Sandstone. The Aztec Sandstone is equivalent to the Navajo Sandstone, which forms the cliffs at Zion National Park. The cliff and hogback exposures between 12:00 and 3:30 consist of Permian Kaibab-Toroweap limestones resting on sandstone and shale about 1,200 feet (365 m) thick. These rocks, in turn, rest on the Bird Spring Formation (fig. 29). At this locality, according to Longwell et al. (1965), the basal 300 feet (90 m) of the sequence tentatively may be assigned to the Permian Queantoweap Sandstone, the middle 800 feet (245 m) of red siltstone and shale may be placed in the Permian Supai Formation (Group), and the uppermost 100 feet (30 m) of sandstone may belong to the Coconino Sandstone — also Permian. About 3,300 feet (1,000 m) of Triassic rocks, Moenkopi Formation (2,300 ft (700 m)) and Chinle Formation (1,000 ft (305 m)), succeed the Kaibab-Toroweap Formations, but are not visible at this point. The Aztec Sandstone rests on the Chinle Formation; thus, the outcrops of Aztec Sandstone visible here are separated from the older sequence by a fault of substantial displacement. The dip slope at the top of the Bird Spring formation is well displayed at 3:30-5:30.

Figure 36. View to northeast along line of traverse of Apex stratigraphic section. 1 – goniatite locality; 2 – approximately at contact between Indian Springs Formation and Bird Spring Formation; 3– Mississippian-Pennsylvanian contact; 4 – approximate position of *Tschussovskenia* (?) biostrome.

STOP 4 — APEX SECTION
SE¼, NE¼, sec. 5, T.19 S., R. 63 E.
Clark County, Nevada

This section (figs. 36, 37), Webster's (1969) Apex Section, is on the gently northwestward-dipping east limb of a syncline in the southern end of the Arrow Canyon Range. The Carboniferous section contains a transitional but predominantly shelf facies of the Monte Cristo Group overlain by the Battleship Wash Formation and basin facies of the Indian Springs and Bird Spring Formations. Only the upper part of the Monte Cristo Group through basal part of the Bird Spring Formation will be examined.

The Battleship Wash Formation (fig. 37) is thin and yields a few poorly preserved megafossils and moderately abundant conodonts of the *Kladognathus-Cavusgnathus naviculus* Zone. It is overlain by 132 feet (40 m) of interbedded gray to black shale, thin ferruginous limestone, orthoquartzite and limestone conglomerate in the Indian Springs Formation. The carbonate content increases upward. Conodonts from the lower 121 feet (37 m) belong to the late Chesterian *Adetognathus unicornis* Zone. The upper 11 feet (3 m) are covered. At 36 feet (11 m) above the base, unit 8 of Webster (1969), a limestone lens contains a cephalopod fauna including *Syngastrioceras walkeri, Cravenocerus* sp., *Pronorites* sp., *Liroceras liratum* and *Mooreoceras* sp. (Webster and Lane, 1967; Webster, 1969). This fauna is of E_2 age and was the first reported occurrence of *Syngastrioceras* from Late Mississippian strata in North America (Webster and Lane, 1967). Lane (1961) reported *Archimedes* from thin limestone beds in the upper part of the Indian Springs at this locality.

The overlying Bird Spring Formation (fig. 37) is dominated by carbonates, moderately to abundantly fossiliferous, and at the center of the syncline has been eroded down to strata of Desmoinesian or Missourian age. The Mississippian-Pennsylvanian boundary occurs at a slight reentrant in the lower part of the first cliff above the Indian Springs Formation (fig. 37). Basal limestones of the Bird Spring formation, unit 18 of Webster (1969), yield conodonts of the *Rhachistognathus muricatus* Zone and the brachiopod naming the *Rhipidomella nevadensis* Assemblage Zone. Immediately overlying limestones contain conodonts of the *Rhachistognathus primus* Zone. A coralline biostrome 217 feet (66 m) above the base contains *Tschussovskenia*? in the lower part of unit 30, Webster (1969). Unit 30 is at the top of the second prominent cliff

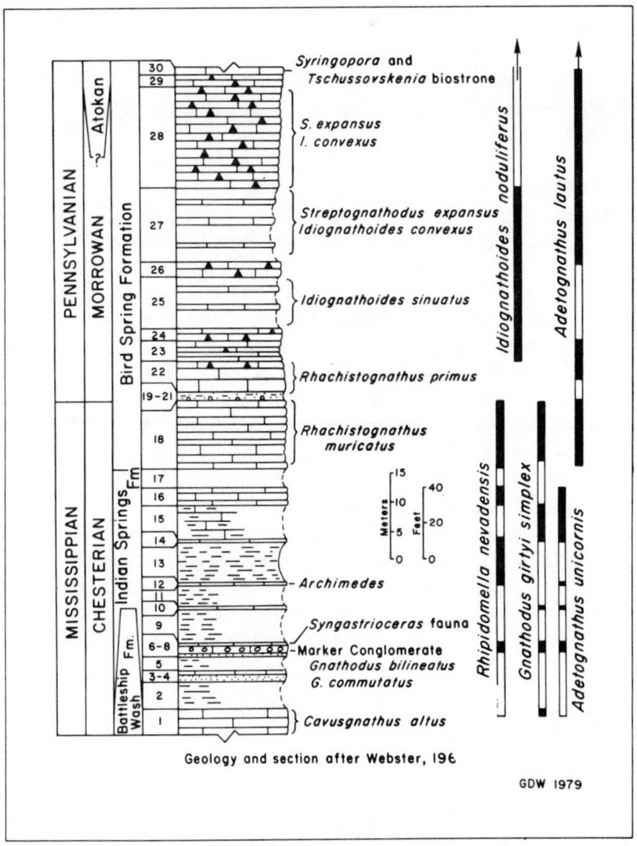

Figure 37. Columnar section of Battleship Wash Formation through basal part of Bird Spring Formation at Apex, Stop 4.

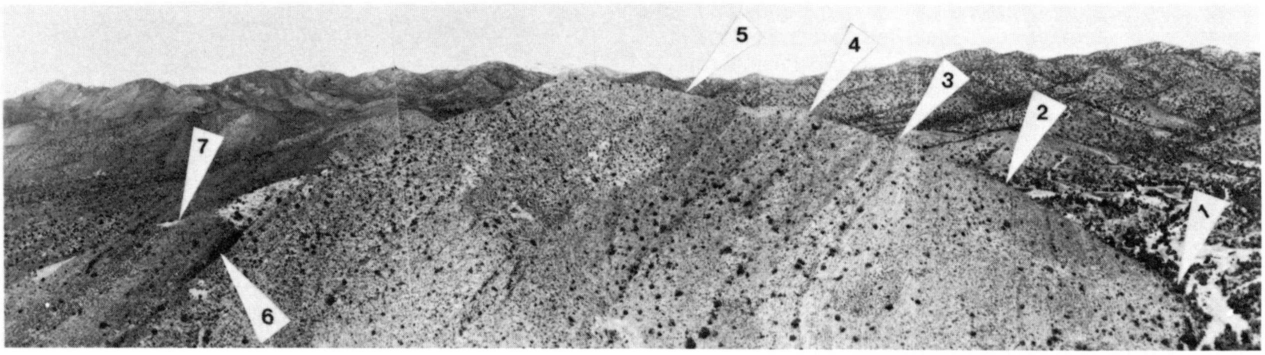

Figure 38. Panoramic view of Devonian and Mississippian portion of Mountain Springs Pass section. 1 – approximate top of Goodsprings Dolomite; 2 – approximate top of Valentine Dolomite Member of Sultan Limestone; 3 – top of Crystal Pass Limestone member of Sultan Limestone; 4 – top of Dawn Limestone of Monte Cristo Group; 5 – top of Anchor Limestone; 6 – outcrop of Arrowhead Limestone; 7 – top of Yellowpine Limestone of Monte Cristo Group. White "blotch" above "6" is an area of secondary dolomitization and mineralization in Yellowpine Limestone.

above the base of the Bird Spring (fig. 36). The alga *Osagia*? occurs 332 feet (100 m) above the base, unit 35 of Webster (1969), and *Chaetetes* occurs in three biostromes between 530 and 702 feet (161 and 213 m) above the base, units 54, 65 and 67 of Webster (1969).

The presence of *Osagia*?, general thinness of the Morrowan and Atokan part of the Bird Spring, abundance of quartz in the limestones, and absence of the "Marker Limestone," which is present to the north and west, suggest that the Bird Spring Formation here is transitional between platform and basin facies. It is distinctly more basinal than platform in aspect.

STOP 5 — KEYSTONE THRUST
SW¼, sec. 26, T. 22 S., R. 58 E.
Clark County, Nevada

Potosi Mountain, at 12 o'clock, is mostly composed of Monte Cristo Group in repeated fault blocks as seen from this point. Aztec Sandstone cliffs abut Devonian and Mississippian carbonates in fault contact at 12:30. To the north, roughly at 3:30, Moenkopi and Chinle outcrops underlie a strike valley extending toward La Madre Mountain, where the trace of the Keystone Thrust is well displayed. Here diverse Paleozoic rocks rest in thrust contact on the Aztec Sandstone. Carboniferous rocks of the Bird Spring Formation and Monte Cristo Group form the skyline on La Madre Mountain. The prominent white scars on dip slopes rising to the skyline at 4 to 5 o'clock are gypsum quarries in the upper part of the Kaibab Formation. These mines, at Blue Diamond, have been active since 1925 and, in 1965, were reported to produce more than 300,000 tons annually (Longwell et al., 1965). The Permian Kaibab and Toroweap Formations here are in stratigraphic continuity with the Jurassic and Triassic rocks below the Keystone Thrust. On the south side of the road, dip slopes descending toward us at 6:00 to 9:30 are at the top of the Kaibab Limestone. Limestone in the Moenkopi Formation forms the stratigraphically highest hogback at 9:30. These rocks are separated from an isolated fault block of Bird Spring Formation to the north along the base of Potosi Mountain.

STOP 6 — MOUNTAIN SPRINGS PASS
NE¼, sec. 19, T. 22 S., R. 58 E.
Clark County, Nevada

At this stop we shall examine a small part of a complete, unfaulted sequence from the base of the Sultan Limestone through the Kaibab Limestone (figs. 38, 39). Principal attention, however, will be paid to the fine exposure of the Monte Cristo Group (fig. 40) as mapped by Hewett (1931) at the time he defined the Monte Cristo Formation and its constituent members. This section is probably the least disturbed complete section in Hewett's map area and has the added attraction of being readily accessible. For these reasons, Langenheim (1956) used it as a standard of reference for the Monte Cristo Group.

After leaving the vehicles we will climb the southeastern flank of the ridge north of the road, passing through the top of the Goodsprings Dolomite, the Ironside Member of the Sultan Limestone, and the Valentine Member of the Sultan Limestone. Hewett describes the Goodsprings Dolomite in very general terms as dolomite of diverse character and color (Hewett, 1931). At this locality, the uppermost, light-tan, rubbly dolomite at the toe of the slope probably belongs to Gans' (1974) Mountain Springs Formation, an equivalent of the Ely Springs Dolomite within the Goodsprings Formation. Hewett (1931) describes the Ironside Dolomite Member as consisting of dark-gray to black dolomite in beds 2 to 5 feet (0.7–1.5 m) thick which combine to form a massive cliff-forming member. He also notes layers of silicified stromatoporoids in the Ironside Member. The overlying Valentine Limestone Member is described as being principally composed of dark-gray to black limestone with interbedded dolomite and layers and lenses of sandstone. These two members combine to form a predominantly dark-colored unit between the underlying Mountain Springs Formation and the Crystal Pass Limestone above. The sequence has not been reexamined in detail since Hewett's (1931) work, but it appears likely that the two members may be roughly equivalent to the upper and lower members of the Arrow Canyon Formation as discussed by Waines (1961), Ryan and Langenheim (1973), and Reinbold and Langenheim (1977). The entire Sultan Limestone, however, is only about one-fourth as thick as the Arrow Canyon Formation, and it is difficult to satisfactorily relate the Ironside Dolomite and Valentine Limestone with the miogeosynclinal sequence in the Arrow Canyon Range.

The Crystal Pass Limestone, however, is a distinctive unit of more or less uniform thickness throughout the field trip area. It has been recognized in an elongate area between Mountain Springs Pass and the Arrow Canyon Range and in the Muddy Mountains and at Arrolime (Langenheim, 1963). The rocks of the Crystal Pass Limestone are an exceedingly uniform pelletoidal limestone that weathers to a flat white or light-gray surface (Carss and Carozzi, 1964) with a few sandy layers intercalated near the base and top. Its basal contact with the Valentine Limestone or with the uppermost Arrow Canyon Formation is gradational in detail, with the contact being taken at the change in color from dark-gray or black to light-gray to

Figure 39. Panoramic view of uppermost portion of Mountain Springs Pass section and Permo-Carboniferous sequence above measured section. 6 – only outcrop of Arrowhead Limestone in vicinity of measured section; 7 – top of Yellowpine Limestone; 8 – approximately at top of Indian Springs Formation; 9 – contact with redbeds of Supai Formation and Toroweap Formation; 10 – cliffs of Kaibab Limestone. Finely layered sequence between "8" and the wash are outcrops of Bird Spring Formation ranging from Late Mississippian into Wolfcampian. Contact between cyclic limestone of Bird Spring Formation and redbeds of Supai Formation is covered by alluvium in wash. 11 – crest of Spring Mountains.

white. A persistent sequence of interbedded sandstone beds just below the base of the Crystal Pass Limestone, however, tends to mark the contact (Ryan and Langenheim, 1973; Reinbold and Langenheim, 1977; Poole, Sandberg, and Boucot, 1977). The upper contact of the Crystal Pass Limestone is an abrupt change from light-colored pelletoidal limestone to black, crinoidal limestone, and, although it may be difficult to find in the field, there is no gradation in lithology or interbedding at the formational boundary.

The Dawn Limestone consists of fine- to medium-grained black limestone that weathers to dark gray. The rock is abundantly fossiliferous, containing spiriferid brachiopods, solitary rugose corals, and *Syringopora* spp. and much of it is crinoidal. Chert is present, but not an obvious component of the rock. The formation tends to weather to form ledgy slopes.

The Anchor Limestone is characterized by thin, nodular beds of black calcilutite interbedded with nodular black chert. Bedding is generally less than one foot thick, except for scattered layers of fine- to medium-grained limestone similar to that of the Dawn Limestone. Fossils are most abundant and consist of spiriferid brachiopods, solitary rugose corals, bryozoans, and echinoderms. The unit weathers to form generally smooth slopes and benches.

The Bullion Formation is characterized by extremely thick and massive bedding and tends to form high, prominent cliffs in outcrop. The rock is relatively coarse-grained limestone, chert is notably absent, and fossils are rare. The Bullion Formation is dolomitic at Mountain Springs Pass, Frenchman Mountain, and Arrolime, among the sections we will visit or be near. It is not dolomitized in the Arrow Canyon Range, Mormon Mountain, or Muddy Mountains, but tends to have local dolomitized masses of secondary origin in these areas.

The Arrowhead Limestone is an extremely persistent marker bed, being recognized in the Goodsprings District, at Arrolime, in the Muddy and Mormon Mountains, Virgin Mountains, and Beaverdam Mountains (Langenheim, 1956, 1963). It consists of thin-bedded, nodular, fine-grained argillaceous limestone and weathers purplish gray on bedding-plane partings and black on fractured surfaces. The limestone contains a distinctive assemblage of chonetid brachiopods and *Straparollus* sp. On weathering it forms a deep reentrant or debris-covered bench and is easily overlooked on the outcrop. The single exposure on the line of traverse at this locality is the only exposure in the vicinity and was unnoticed by Hewett (1931).

The Yellowpine Limestone is characterized by beds from 2 to 4 feet (0.7–1.5 m) thick and generally is topographically expressed as part of the Bullion Formation cliff. The limestone is crinoidal and generally darker colored than the Bullion Formation. Abundant fossils characterize the Yellowpine, especially *Syringopora* sp. and both solitary and compound rugose corals. Chert is largely absent in the Yellowpine Limestone, but there is a well-defined thin chert layer near the top in most of the field-trip area.

The Battleship Wash Formation is not present at Mountain Springs Pass, in contrast to the Apex Section, northeast of the Las Vegas Shear at Stop 4. Thus, the Mountain Springs Pass Section is a platform sequence, in this respect, suggesting that it has been moved northwest relative to Frenchman Mountain and Apex by movement on the Las Vegas Shear.

There has been no significant paleontologic work done on the Monte Cristo Group at Mountain Springs Pass or in the near vicinity since Girty identified Hewett's (1931) fossils and assigned the Dawn, Anchor and Bullion Limestones to the Early Mississippian, the Arrowhead Limestone to the Middle Mississippian, and the Yellowpine Limestone, probably, to the Middle Mississippian. These assignments are more or less compatible with Pierce and Langenheim's (1972) assignment of the Dawn and lower Anchor to the Kinderhookian, the upper part of the Anchor and Bullion to the lower Valmeyeran (through equivalents of the lowermost part of the Upper Burlington), and the Yellowpine Limestone to Upper Burlington through Salem equivalents in the Valmeyeran. The ages of the Monte Cristo Group formations in the more geosynclinal Arrow Canyon Section, however, may or may not be the same as their ages here in a platform sequence. We will see part of the sequence studied by Pierce and Langenheim (1972) in the Arrow Canyon Range tomorrow, but will not visit their measured section.

Conodonts recovered by Webster from limestone beds in the poorly exposed Indian Springs Formation at this locality include *Adetognathus unicornis, Gnathodus bilineatus, G. commutatus, G. girtyi simplex, Spathognathodus rexroadi,* and *Rhachistognathus muricatus.* Both the *Adetognathus unicornis* and *Rachistognathus muricatus* Zones fo Late Chesterian age are represented; this is the southwesternmost known locality of these zones in the Great Basin.

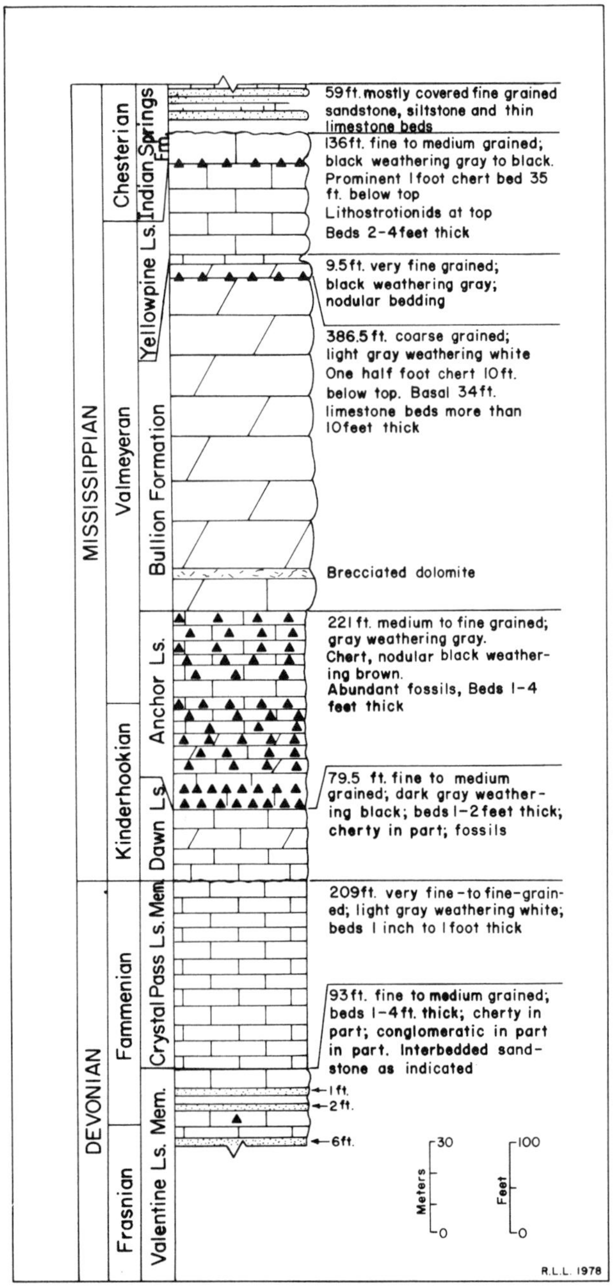

Figure 40. Columnar section of Late Devonian into Late Mississippian strata at Mountain Springs Pass. Sultan Limestone is represented by upper part of Valentine Limestone Member and Crystal Pass Limestone Member. Monte Cristo group includes all strata from the Dawn through Yellowpine Limestones with Arrowhead Limestone (9.5–ft-(3m) thick unit immediately below Yellowpine) unlabeled.

ROAD LOG—SEVENTH DAY: CLARK COUNTY, NEVADA

Ralph L. Langenheim, Jr.[1] and Gary D. Webster[2]

The road log for the seventh day's excursion in Clark County, Nevada, begins at the intersection of Fremont Street and Las Vegas Boulevard in downtown Las Vegas (figs. 22, 23). We proceed northerly on U.S. 95 to the interchange with Interstate 15, at which point we enter the Interstate, to travel northeasterly to Glendale, Nevada. En route we will discuss the regional geology along the thrust belt separating the geosyncline and platform and will see extensive outcrops of Carboniferous and other Paleozoic rocks. On arrival at Glendale, we leave I–15, traveling northwest on Nevada Highway 7 and secondary roads to reach Arrow Canyon. Arrow Canyon is a narrow gorge, superposed across the northern end of the Arrow Canyon Range, in which most of the Monte Cristo Group, the Battleship Wash Formation, the Indian Springs Formation, the entire Pennsylvanian part of the Bird Spring Formation, and most of the Wolfcampian part of the Bird Spring Formation are exposed in a steeply inclined, but structually intact, section that is essentially completely exposed. This readily accessible stratigraphic section has been repeatedly studied and is the best known Late Mississippian through Early Permian sequence in the southern part of the Cordilleran Miogeosyncline. We will spend most of the day examining this locality in some detail. After leaving the Arrow Canyon section, we proceed around the north and west sides of the Arrow Canyon Range via Nevada 7 and U.S. 93, affording an excellent view of the complete Late Cambrian through Mississippian sequence exposed on the west face of the Arrow Canyon Range, another well-studied reference section for the region. We will then return to Las Vegas via I–15.

Total Mileage

0.0 Log begins at intersection of Fremont Street and Las Vegas Boulevard. Proceed northeast on U.S. 93–95, Las Vegas Boulevard.

0.3

0.3 Bear left on U.S. 95 to freeway interchange.

0.5

0.8 Bear right on access ramp leading to I–15; following sign to Salt Lake City.

0.3

1.1 Enter freeway, traveling northeasterly. Gass Peak, southern end of Las Vegas Range, southern end of Arrow Canyon Range, and southern end of Dry Lake Range in clockwise order from 9:30 to 11:30. Almost all rocks visible in these ranges from this point are in Bird Spring Formation. Frenchman Mountain and Sunrise Mountain are at 1:30.

4.6

5.7 Note well-developed alluvial fans to left of road. A smaller fan at 9:30 heads against Gass Peak and a larger fan at 10 o'clock heads against south end of Las Vegas Range. Webster's (1969) Gunnery Range section is to right of this fan and his Gass Peak section is to left. Gass Peak itself is at 9:30 and is highest point, at 6,943 feet (2,111 m), on horizon in this view.

3.1

8.8 Curve in road. Higher level alluvial fan surfaces, capped with caliche, cap remnant hills along road. Muddy Peak is at about 1:30.

2.4

11.2 Low hills on skyline at 12 o'clock are in southern end of Dry Lake Range. High peak at 1:30 on skyline is Muddy Peak. Sunrise Mountain is at 3 o'clock. Mountains in far distance are Black Mountains on Arizona side of Lake Mead.

5.0

16.2 Exit 58 and overpass, old U.S. 91–93 to Apex. Turn here if you wish to visit Stop 4, Apex Section, of Road Log—Sixth Day: Clark County, Nevada. Proceed straight ahead on interstate for today's trip. Pink sediments at 11:30–12:30 are Pliocene Muddy Creek Formation, valley fill that very nearly covered much of present mountain ranges before Grand Canyon began downcutting.

1.7

17.9 Arrolime Quarry at 10:30. Principal cut is in Late Devonian, (Fammenian) Crystal Pass Limestone and in uppermost Arrow Canyon Formation. Arrow Canyon Formation is roughly equivalent to Sultan Limestone of Spring Mountains and is of Late Devonian (Frasnian and early Fammenian) age. Rounded ridge to left is in Bullion Formation of Monte Cristo Group. Langenheim's Apex section is conformable with these rocks to north of Arrolime quarry. Bird Spring Formation exposed to right of road is in lower block of Dry Lake Thrust, which crops out along southeastern side of summit between Arrow Canyon and Dry Lake Ranges at this point. As of 1965 (Longwell et al., 1965), production of high-calcium limestone (98.05% $CaCO_3$) averaged more than 500,000 tons annually from about 200 feet of productive Crystal Pass Limestone.

0.7

18.6 Underpass, spur line of Union Pacific Railroad crossing interstate.

2.3

20.9 Underpass, main line of Union Pacific Railroad crossing interstate.

1.8

22.7 Interchange; U.S. 93 diverges to northwest; continue straight ahead.

0.3

23.0 Pull off to right for orientation.

[1] Department of Geology, University of Illinois, Urbana, Illinois 61801.
[2] Department of Geology, Washington State University, Pullman, Washington 99163.

STOP 1—DRY LAKE VALLEY. Discussion of geography and geology of the thrust belt northeast of Las Vegas.

4.8

27.8 Overpass; I–15 crosses main line of Union Pacific Railroad. At pulloff just beyond overpass we are very nearly on trace of Dry Lake Thrust. Crest of Sheep Range visible beyond Arrow Canyon Range at 8:30–9:30. That part of Arrow Canyon Range at 10 o'clock is chiefly Monte Cristo Group as seen from this point, but much Devonian rock, mostly Arrow Canyon Formation, also is visible. Monte Cristo Group outcrops are generally lighter gray and characterized by imposing cliffs of Bullion Formation, here a limestone. Outcrops of Arrow Canyon Formation are darker gray, less homogeneous, having a banded appearance. Relatively thin Crystal Pass Limestone is readily distinguished by its prominent, very light-gray outcrop. Flat-topped hill on right-hand portion of main mass of Arrow Canyon Range is composed of Arrow Canyon Formation and is on axis of Arrow Canyon Syncline of Longwell et al. (1965). This mountain and syncline will be seen from opposite side after we leave Arrow Canyon later in the day. Base of flat-topped synclinal hill is composed of Laketown Dolomite (Silurian), Ely Springs Formation (Middle and Late Ordovician), Eureka Quartzite (Middle Ordovician), and Pogonip Group (Early Ordovician) that is in stratigraphic and structural continuity with Nopah Formation (Late Cambrian) resting on Dry Lake Thrust in its northeastern exposure. Lower hills at right-hand margin of Arrow Canyon Range, made of layers dipping northwestward, are in overturned sequence ranging from Pogonip Group through Monte Cristo Group in an overturned syncline below Dry Lake Thrust. Arrow Canyon Section, Stop 3 of today's trip, is at northern end of this structural block. As we leave this point we will cross northern end of Dry Lake Range, traversing much of Pennsylvanian part of Bird Spring Formation. Careful observation of lower bedrock slopes to right of road, depending on lighting, may provide a view of a recent fault scarp bounding Dry Lake Range.

2.4

30.2 Leave Dry Lake Range; California Wash to right. Note that Dry Lake Range is in structural continuity with overturned block along east side of Arrow Canyon Range and, presumably, is bordered along its northwestern side by the Dry Lake Thrust, here buried by alluvium, and, probably, Muddy Creek Formation.

1.9

32.1 Northern end of North Muddy Mountains visible on skyline at 12 o'clock. Arrowhead Fault, a major transverse fault cutting across dominant Basin and Range fault sets, separates North Muddy Mountains from Muddy Mountains proper at gap on skyline at 2 o'clock. Muddy Peak, highest point in Muddy Mountains at 5,432 feet (1,651 m), is sharply pointed crest at 3 o'clock.

North Muddy Mountains are almost entirely composed of parrallel folds of Permian and Mesozoic rocks, in contrast to Muddy Mountains. Muddy Mountains expose Cambrian through Permian Paleozoic rocks, are predominantly of Devonian through Pennsylvanian rocks, are overlapped and partially buried by diverse Tertiary sedimentary rocks, and their structure is dominated by Muddy Mountains Thrust and Basin and Range tensional faulting. Pink outcropping low on slopes below lowpoint along crest of Muddy Mountains is an exposure of Late Triassic or Early Jurassic Aztec Sandstone in Buffington Pockets, fensters exposing lower plate of Muddy Mountains Thrust.

Longwell (1928, 1949, 1952, 1962) devoted a good portion of a long and productive research career to describing and interpreting the geology of the Muddy Mountains and North Muddy Mountains. As a consequence, they have become a classic area, typifying the structure of the southern part of the thrust belt separating the Cordilleran Miogeosyncline from the adjacent shelf.

Pliocene Muddy Creek Formation forms pink outcrops along road from here to Arrow Canyon stop. Mesas and scarps in which Muddy Creek Formation is exposed are capped by an extensive caliche crust. This crust preserves large areas of old fan surfaces predating downcutting caused by development of external Colorado River drainage in this area. Old fan surface is estimated by Gardner (1972a) as being somewhere between about 900,000 and 2,600,000 years old.

1.9

34.0 Exit 75, Crystal; Nevada 40 to south, reaching Overton via Valley of Fire State Park. A now-defunct gasoline station at this junction was the excuse for a place name at this spot. Nevada 40 is paved and passable essentially at all times. The Valley of Fire State Park has camping facilities, drinking water, and a visitor center and museum. The park affords excellent exposures of the Arrowhead Fault scarp and of Triassic and Jurassic redbeds faulted against Paleozoic carbonate rocks to south. Caliche caps old fan surface above exposed Muddy Creek Formation to left at this point.

0.7

34.7 Buffington Pockets clearly visible at 4 o'clock, given good atmospheric and light conditions. Beaverdam Mountains, extending into Utah and Arizona, visible at 12 o'clock on skyline.

1.4

36.1 Highest peak 6,205 feet (1,886 m) in southern end of Mormon Mountains at 11:30. This peak is a strike ridge of Monte Cristo Group and is quite spectacular, seen end on, as at Arrow Canyon later in the day.

1.3

37.4 Caliche-capped mesas of Muddy Creek Formation to right along road ahead.

1.6

39.0 Ute interchange; secondary roads lead northward to Ute Siding on the Union Pacific Railroad and onward to a microwave relay and an air navigation radio beacon on crest of southeasternmost ridge of Arrow Canyon Range. These roads are maintained and passable at almost any time in a standard sedan. In addition, a graded road, unmaintained since about 1950, continues northwesterly to the Tiffany Quarry. This quarry produced something less than 2,000 tons of quartzite annually during intermittent operations prior to 1954 (Longwell et al., 1965). This is the Silica Quarry locality of Langenheim et al. (1962) and is the area of numerous published observations on the Pogonip Group, Eureka Quartzite, Ely Springs Dolomite, Piute Formation, Moapa Formation, and Arrow Canyon Formation, all of which are well exposed in readily accessible outcrops in the immediate vicinity of the Tiffany Quarry. Travel beyond the end of the maintained segment of the road, however, should be conducted with extreme care as it is a long way out to the highway and there is no water or shelter in the area.

0.7

39.7 Road passes through well-exposed Muddy Creek Formation as we cross main drainage of California Wash.

3.1

42.8 Byron exit: stacks at 10:30 are at Moapa powerplant of the Nevada Power Company. This plant, opened in the 1960's to help carry the ever-increasing electrical demand of Las Vegas and vicinity, burns Cretaceous coal from the Book Cliffs of Utah to make steam from ground water pumped from the head of the Moapa Valley at the mouth of Arrow Canyon. The coal is hauled more than 300 miles by rail from the mines to the powerplant.

3.4

46.2 Rest stop at right; pull off to right and stop.

STOP 2 — GLENDALE THRUST. Prominent, very light-gray outcrop band on slopes of middle ridgeline at 8 o'clock is outcropping Crystal Pass Limestone at top of Devonian and is within an overturned sequence below Dry Lake Thrust Fault. Bird Spring Formation comprises the lower ridge in front and extends beyond the Crystal Pass Limestone outcrops to right and left. Monte Cristo Group, which includes a wider, but less distinct, light-gray outcrop band in Bullion Limestone, forms generally darker lower slopes of middle ridge. Skyline ridge, as seen from here mostly exposing dip slopes of Arrow Canyon Formation, is main ridge of Arrow Canyon Range. We will pass southwesterly along the valley beyond this ridge as we return to Las Vegas later today. Power plant at Moapa stands directly in front of low pass or saddle separating Arrow Canyon Range from Meadow Valley Range at 9 o'clock. Skyline beyond saddle is crest of Sheep Range. Meadow Valley Wash, at 10 o'clock, lies between Meadow Valley Range and Mormon Mountains at 11:30. Meadow Valley Range is a northward extension of Arrow Canyon Range, eastern foothills and Dry Lake Range structural block, but is broken by a transverse fault a few miles north of its southern end. It is covered at northern end by Eocene to Oligocene ash flow tuffs and associated volcanic rocks. Mormon Mountains are principally composed of Paleozoic sedimentary rocks and are a thrust-faulted extension of Muddy Mountain structural belt.

Caliche-capped, older alluvial fan surfaces apron Mormon Mountains and North Muddy Mountains to right of road. These fan surfaces are incised by Meadow Valley Wash, the Muddy River, and their tributaries, thus exposing pink scarps of Muddy Creek Formation in front of Mormon Mountains and around Moapa as viewed from this point.

0.9

47.1 Hidden Valley exit, continue straight ahead.

1.3

48.4 Roadcuts in Permian Kaibab Limestone on either side of road as we descend into Muddy Valley. Pink, earthy exposures are Muddy Creek Formation.

1.0

49.4 Turnoff to right for Moapa and Glendale; leave I–15 at this point, being careful of the tight turn. Go slowly.

0.3

49.7 Stop sign; proceed straight ahead on Nevada 7. Check your gasoline supply. We will travel about 75 to 80 miles before our return to Las Vegas, where the first gasoline stations will be encountered. If you are leaving the log route to travel northward toward Pioche, Caliente, Tonopah, Ely, or elsewhere, the first services will be found at Alamo, also about 75 miles, via our log, from this point. There are no habitations or water supplies, other than Arrolime Quarry, beyond the mouth of Arrow Canyon on the Las Vegas route.

Cross Muddy River just beyond this intersection. Muddy River is a permanent stream rising in springs a few miles northwest.

0.9

50.6 Coarse-grained alluvium in roadcut. This Quaternary material is much younger than Pliocene to Early Pleistocene Muddy Creek Formation.

0.8

51.4 Road reaches level of caliche-armored, older alluvial surface. Most of our traverse for next six miles will be on this surface, estimated to be between about 1 to 2.6 m.y. B.P. or Early Pleistocene by Gardner (1972a).

1.1

52.5 Turnoff to left for Moapa, followed directly by overpass crossing mainline of Union Pacific Railroad.

0.2

52.7 Turnoff to left for Moapa post office.

0.7

53.4 We are now traveling on an extensive segment of well-preserved Early Pleistocene alluvial surface. Note thin soil and projecting caliche; this stuff "kills" tires quickly.

0.5

53.9 Turnoff to left to enter Moapa Indian Reservation; continue straight ahead. The Moapa band is a small group of Paiute Indians, living to the southwest on irrigated land. These people are very probably the direct descendents of the Indians responsible for some of the petroglyphs in Arrow Canyon and, according to local informants, still conduct some ceremonies in the canyon. Meadow Valley Range visible ahead to right of road; Arrow Canyon Range to left. High, skyline crest beyond is in Sheep Range, here composed of Cambrian and Ordovician rocks.

3.0

56.9 Warmsprings turnoff to left; continue straight ahead. These are the hot springs giving rise to the Muddy River. The palms are *Washingtonia filifera,* native to the Salton Basin and its bordering mountains in southernmost California. They are cultivated here. Proceed northwesterly in irrigated upper Moapa Valley, with well-exposed, pink slopes of Muddy Creek Formation scattered on all sides.

3.7

60.6 Junction with paved road; turn left, leaving Nevada 7 for access to Arrow Canyon. Beware of sharp dip crossing Arrow Canyon Wash. Do not enter dip during flash floods!

0.4

Figure 41. Virgilian-Wolfcampian boundary as exposed on north side of Arrow Canyon. Boundary is at top of ledge marked "1" on photograph and lies between units 224 and 225 of Cassity and Langenheim (1966). At this time a large red "39" is painted on top of this ledge and a red "38" is painted on unit 222 of Cassity and Langenheim (1966).

Figure 42. Units 217 through 226 of Cassity and Langenheim (1966) as exposed on north side of Arrow Canyon. 1 – unit 217, marked in field by a large red "37." 2 – Virgilian-Wolfcampian boundary between units 224 and 225 of Cassity and Langenheim (1966). 3 – approximate base of dolomite of unit 227 of Cassity and Langenheim (1966).

61.0 Primitive desert road leaving to right into Arrow Canyon. Turn right and proceed through refuse dump. Widespread dark-colored, stony surface in flat area here is formed by desert varnished, lag gravel covering silt. This is referred to as desert pavement.

0.3

61.3 Prudent drivers in cars with low clearance should carefully investigate the crossing ahead. The wash may be eroded so that your car will strand or there may be water in the wash. On the other hand, the road may be passable for an indeterminant distance up the wash and generally is easily driven in "pickup" type vehicles. Four-wheel-drive vehicles are able to traverse the entire section at almost any time, being blocked only when the wash is in flood. The wash, incidentally, is hazardous if flash flooding is likely. This is the last easy turnaround point on the section traverse that is below any hazard.

1.0

62.3 Prominent cliffs in thick caliche at this point and before. Bedrock here is in early Wolfcampian, and cliff to right is a relatively persistent marker bed. False summit above is in so-called "coral bed," containing abundant *Durhaminia* sp. This coral and its associates form a very useful "key bed" for correlation in the Great Basin. At this point, we also approach an abrupt flexure where Bird Spring Formation is bent upward to west beneath Dry Lake Thrust trend. The fault, however, has not yet been traced this far northward and, presumably, is buried beneath alluvium and Muddy Creek Formation west of upper end of Arrow Canyon. Note nearly vertical trace of bedding on skyline at 2:30, which is in Permian part of Bird Spring Formation. Pennsylvanian rocks crop out ahead at 12 o'clock, but cliffs to left at 10 o'clock are in Permian.

0.3

62.6 Enter wash. Axis of syncline at 9 o'clock on skyline. Proceed in wash for a short distance and again climb out on north side.

0.2

62.8 **STOP 3—ARROW CANYON.** From this point we will ascend the wash, gradually descending stratigraphically through Pennsylvanian portion of Bird Spring Formation, Mississippian portion of Bird Spring Formation, Indian Springs Formation, Battleship Wash Formation, and part of Monte Cristo Group. All notations from this point onward, until we retrace our steps, will refer to parts of Stop 3.

At this point we are at the Pennsylvanian-Permian boundary, which may be identified by reference to Figures 41 and 42. In Figure 41, G. D. Webster has his hand on the top of unit 224 of Cassity and Langenheim (1966), identified as top of the Pennsylvanian (Virgilian) and base of the Permian (Wolfcampian). This unit contains *Triticites ventricosus sacramentoensis* and overlying unit 225 contains *T. creekensis* (Cassity and Langenheim, 1966). Langenheim and Langenheim (1965) report *T.* cf. *T. californicus* and *Schwagerina* sp. in unit 223, about 18 feet below top of unit 224. These earlier reports were not verified by Cassity and Langenheim (1966) because the presumed *Schwagerina* sp. was not recollected and the *T.* cf. *T. californicus,* upon examination of a larger collection, was reclassified as *T. creekensis*.

0.05

62.85 Enter wash at Missourian-Virgilian boundary. Base of Virgilian is at base of upper, massive ledge on spur meeting road at its entrance into wash (figs. 43, 44). This is about one-fourth of the distance upward from base of unit 211 of Cassity and Langenheim (1966) and Langenheim and Langenheim (1965). Cassity and Langenheim (1966) report the following fusulinids at or near this point in section: *Pseudofusulinella utahensis* in both Missourian and Virgilian parts of unit 211 and in unit 213; *Triticites birdspringensis* in Virgilian part of unit 211 along with *T. muddiensis* which also occurs in units 212, 213, 215, and 217; *Triticites* cf. *T. collus* occurs in unit 209 and *T. ohioensis* (?) in Missourian portion of unit 211.

0.05

Figure 43. 1 – Missourian-Virgilian boundary as exposed on north side of Arrow Canyon in the middle of unit 211 of Cassity and Langenheim (1966). Bed 210 is marked in field by a large red "35" and bed 211, forming point of spur entering canyon, is painted with a large red "36."

Figure 44. Entire Virgilian sequence exposed on north side of Arrow Canyon. 1 – base, 2 – top, of Virgilian. 3 – ledge containing unit 217. Virgilian comprises units 211 through 224 of Cassity and Langenheim (1966).

Figure 45. Desmoinesian-Missourian boundary as exposed on north wall of Arrow Canyon. 1 – contact between units 192 and 193 of Cassity and Langenheim (1966).

Figure 46. Desmoinesian-Missourian boundary as exposed on north wall of Arrow Canyon. 1 – base, 2 – top, of Cassity and Langenheim's (1966) unit 190. 3 – contact between Cassity and Langenheim's (1966) units 192 and 193, the Desmoinesian-Missourian boundary. 4 – top of Cassity and Langenheim's (1966) unit 193.

62.9 Desmoinesian-Missourian boundary at 3 o'clock across wash (figs. 45, 46). Base of Missourian is at top of Cassity and Langenheim's (1966) and Langenheim and Langenheim's (1965) unit 192. Unit 192 weathers more brown than gray, in contrast to unit 193 which is lighter colored and grayer. Cassity and Langenheim (1966) report the following fusulinids at or near the contact: *Pseudofusulinella logandalensis* and *Oketaella* cf. *O. lenensis* occur in unit 193, defining base of Missourian (figs. 45, 46). There are no fusulinids noted in bed 192, but bed 191 contains *Fusulina retusa* and *Bartramella bartrami* (fig. 46).

Prominent rusty-weathering bed cropping out just below abrupt upstream narrowing of canyon is "Tungsten Gap Chert" of Castle (1967), Micklin (1967) and Smosna (1970). This very distinctive bed of cherty, silty limestone, bed 183 of Cassity and Langenheim (1966), is readily distingushed in the field (fig. 47) and shows up well on aerial photographs. As a consequence it has been a key to geologic mapping in the area. It also occurs within Dry Lake, Las Vegas, Arrow Canyon, and Meadow Valley Ranges, and part of the thick sandstone in upper part of Callville Formation at Frenchman Mountain is believed equivalent to "Tungsten Gap Chert."

0.2

63.1 Atokan-Desmoinesian contact of Webster (1969) is just downstream from wash-crossing as of 1978 (fig. 48). Webster (1969) reports the lowest occurrence of *Fusulina* in the bench between the prominent cliff in the upper right of Figure 48 and the indistinct ledge below. Cassity and Langenheim (1966) and Nelson (1973) place the base of the Desmoinesian approximately 160 feet (48 m) stratigraphically above this point. Cassity and Langenheim (1966) report *Fusulinella devexa* in their bed 150 succeeded by an occurrence of *Wedekindellina* sp. in unit 152. The highest occurrence of *Chaetetes* in the Arrow Canyon Section is in the upper part of the ledge behind the man shown in Figure 48. This is bed 108 of Webster (1969), bed 42 of Nelson (1973), and probably bed 133 of Cassity and Langenheim (1966).

0.1

Figure 47. "Tungsten Gap Chert" as exposed on north wall of Arrow Canyon. 1 – base, 2 – top, of "Tungsten Gap Chert," unit 183 of Cassity and Langenheim (1966).

Figure 48. Desmoinesian-Atokan boundary as exposed on north wall of Arrow Canyon. 1, 2, 3 – base of Nelson's (1973) units 40, 41 and 42, respectively. All three units are within Webster's (1969) unit 107. 4 – top of Webster's (1969) unit 108, Nelson's (1973) unit 42, and, probably, Cassity and Langenheim's (1966) unit 133. This bed contains stratigraphically highest occurrence of *Chaetetes* spp. at Arrow Canyon. 5 – base of Desmoinesian as determined by Webster (1969); 6 – top of basal bed in Desmoinesian. This bed contains lowest occurrence of *Fusulina* spp. in Arrow Canyon section according to Webster (1969) and is about in the middle of Nelson's (1973) unit 43. 7, 8 – base and middle of Nelson's (1973) unit 45.

Figure 49. Contact between lower and upper Atokan as exposed on north side of Arrow Canyon. 1 – base of unit 110 of Cassity and Langenheim (1966). 2 – base of Nelson's (1973) unit 18, Coogan's (1962) unit 162, and Webster's (1969) unit 95. This bed, at base of upper Atokan Zone of *Fusulinella* and *Chaetetes*, is painted with a large orange "26." 3 – base of Cassity and Langenheim's (1966) unit 113, Webster's (1969) unit 96, Nelson's (1973) unit 19 and Coogan's (1962) unit 163. 4 – base of unit 115 of Cassity and Langenheim (1966).

Figure 50. Morrowan-Atokan boundary as exposed on north wall of Arrow Canyon. 1 – base of *Profusulinella* Zone and of Atokan. This is base of Nelson's (1973) unit 2. 2 – base of Nelson's (1973) unit 3, which contains corals. 3 – base of his unit 4. 4 – base of his unit 9.

63.2 *Fusulinella* in outcrops on both sides of canyon. The man has his hand on unit 95 of Webster (1969) (fig. 49), equivalent to unit 110 of Cassity and Langenheim (1966), to unit 162 of Coogan (1962), and to unit 17-18 of Nelson (1973). Lower portion of this ledge contains lowermost *Chaetetes* in Arrow Canyon Section. Upper portion of ledge, unit 18 of Nelson, contains abundant *Fusulinella.* Cliff beyond, behind automobile, includes Nelson's (1973) units 21 through 25 and unit 97 of Webster (1969). *Chaetetes* occurs in Nelson's (1973) unit 25 and in Webster's unit 97. The lowest occurrence of *Chaetetes* defines base of Webster's (1969) *Fusulinella–Chaetetes* zone, Coogan's (1962, 1964) zone of *Fusulinella,* and Cassity and Langenheim's (1966) zone of *Fusulinella,* which are referred to either as upper Atokan or upper Derryan.

0.05

63.25 Bed on north side of canyon contains abundant *Profusulinella* sp. (fig. 50) and is within Webster's (1969) unit 87. It also is part of Cassity and Langenheim's (1966) unit 101. Nelson's (1973) unit 3 forms ledge to right of man in Figure 50 and shaly sequence below; Nelson's unit 2 below contains abundant *Profusulinella spicata* and *Caninia* sp. Nelson's units 6 through 10 form cliffs to right of bench above *Profusulinella*-bearing ledge, with his unit 10 forming point of spur entering flat valley floor. Coral locality is probably Coogan's (1962) unit 142. Both Webster (1969) and Cassity and Langenheim (1966) place base of *Profusulinella* Zone and base of Atokan or Derryan at this point.

0.05

63.3 Here we cross *Paragassizocrinus calyculoides* layer, unit 60 of Webster (1969) (fig. 51). This crinoidal biostrome is a useful marker bed within the eastern portion of the Great Basin, being reported as far north as the

Figure 51. *Paragassizocrinus caliculoides* bed, unit 60 of Webster (1969), as exposed on south wall of Arrow Canyon: 1 – top, 2 – bottom. 3 – top of Webster's (1969) "Marker Limestone."

Figure 52. Mississippian-Pennsylvanian boundary as exposed on north wall of Arrow Canyon (4). 1-8 – base of Webster's (1969) units 25, 26, 27-28, 29, 30, 31, 32 and 33 in order. Units 27 and 28 are not separated in illustration.

Bridal Veil Falls section of Washburn (1968) at Provo and as far south as Spring Mountains. Webster's unit 58, 136.5 feet (41 m) below the *Paragassizocrinus* bed, contains the coral *Tschussovskenia* ? and is a few feet below the base of his "middle" limestone marker bed, a cliff-forming sequence of thin-bedded, very arenaceous, cherty, fine-grained limestone that generally weathers as a light-brown, massive unit. This unit is another key interval for mapping in the Arrow Canyon, Meadow Valley, Dry Lake, and Las Vegas Ranges. Webster's (1969) unit 56, considered equivalent to Cassity and Langenheim's (1966) unit 40 by Lane, Sanderson and Verville (1972), is an additional 64.5 feet (20 m) below the *Tschussovskenia* ? bed. This is the level at which Lane, Sanderson and Verville (1972) place the Morrowan-Atokan boundary. This horizon is probably at the top of unit 126 of Coogan (1962). Coogan (1962), however, places the base of the Derryan another 30 feet (9 m) lower at the base of his unit 125.

0.1

63.4 "Drag-fold" unit well exposed on north side of canyon and contains abundant invertebrate megafossils. This is unit 33 of Cassity and Langenheim (1966) and unit 46 of Webster (1969).

0.1

63.5 Cross wash; Mississippian-Pennsylvanian boundary just below ledge above prominent bench on north side of canyon (fig. 52). Man in figure and claim post are both about at contact. Cliff includes beds 30-32 of Webster (1969), man is standing on bed 29 (approximately) of Webster (1969), and post is about at top of bed 28 of Webster (1969). Top of bed 28 is top of Chesterian portion of Bird Spring Formation.

0.1

63.6 Top of Battleship Wash Formation and point at which "road" enters wash and gorge of Arrow Canyon. This is the last spacious flat spot for parking and turning vehicles and road beyond may or may not be passable for vehicular traffic. There may be a mudhole blocking lower end of gorge; if there is not, remainder of gorge is probably floored by relatively smooth gravel. Do not, under any circumstances, enter the gorge if thunder storms appear likely: this canyon flash-floods.

Figure 53. View northeast across strike valley in Indian Springs Formation from mouth of gorge of Arrow Canyon. 1 – uppermost Battleship Wash Formation; 2 – approximately at top of Indian Springs Formation; 3 – base of Pennsylvanian.

We have just crossed an open, strike valley underlain by Indian Springs Formation. Looking northeast, Mississippian-Pennsylvanian boundary lies in notch at base of thickest, most massive cliff-forming unit in basal Bird Spring Formation (fig. 53). Contorted, fossiliferous limestone layer approximately in middle of Indian Springs strike valley, unit 9 of Webster (1969), is "marker conglomerate" of Webster and Lane (1967) and was seen yesterday at Apex. The similarity of deformation at the two localities is notable. Fossils from Indian Springs Formation and overlying basal Bird Spring Formation at this locality have been described by Lane (1963), and others from this locality have been listed by Webster and Lane (1967) and Webster (1969). Shaly beds, which dominate middle and lower Indian Springs Formation, are largely covered, but one of the best exposures is located just south of wash and may readily be viewed from *Stigmaria wedingtonensis* locality on top of Battleship Wash Formation dip slope. Here stumps of *Stigmaria wedingtonensis*, apparently rooted on topmost bedding plane of Battleship Wash Formation, have been described by Pfefferkorn (1972). Please do not collect or batter these fossils: they are unique to the area.

Figure 54. View toward mouth of gorge, or "narrows," of Arrow Canyon from east. A good outcrop of lower part of Indian Springs Formation is exposed behind shoulder of slope marked "1." 2 – *Stigmaria weddingtonensis* locality; 3 – dip slope at top of Battleship Wash Formation. Base of Meramecian is about 8 feet below the foot of cliff marked "4," and base of Battleship Wash Formation is on bench marked "5." Yellow Pine Limestone and Bullion Formation crop out in gorge at "6."

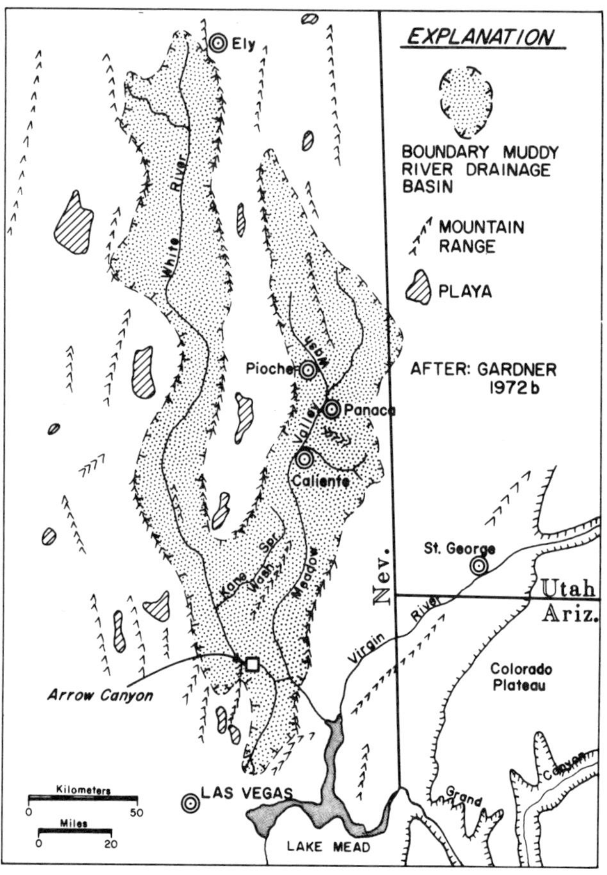

Figure 55. Map of Muddy River – White River – Meadow Valley Wash drainage basin. Modified after Gardner (1972b).

Western wall of Indian Springs strike valley is a dip slope of Battleship Wash Formation (fig. 54), pierced by superposed gorge or "narrows" of Arrow Canyon. Arrow Canyon connects northward with White River Valley which was a through-flowing drainage system during glacial times in the Pleistocene (fig. 55). This valley now contains a series of permanent and intermittent streams, along with lakes and dry lakes. It is more than 250 miles long, rising near Ely, Nevada (Gardner, 1972 b).

Proceed into gorge, noting bench, lithostrotionid biostromes and basal sandstone at contact between Yellowpine Limestone and Battleship Wash Formation (fig. 56). This apparent disconformity was cited as the Chesterian-Meramecian boundary by Langenheim et al. (1962), Langenheim and Langenheim (1965), and Coogan (1962, 1964), but Rice and Langenheim (1974) found that the contact lies wholly within the uppermost Valmeyeran (Meramecian) *Apatognathus scalenus-Cavusgnathus* conodont zone. The series boundary, therefore, is about 40 feet (12 m) above the base of the formation, approximately in the middle of unit 8 of Rice and Langenheim (1974). The top of the Battleship Wash Formation was thought to be at the top of the Chesterian by Langenheim et al. (1962), who placed the Indian Springs or BS_b unit in a questionably Mississippian or Pennsylvanian limbo, and by Langenheim and Langenheim (1965), who placed the Indian Springs Formation in the Springeran. Coogan (1964) placed the Chesterian-Morrowan boundary at the base of the "marker conglomerate" within the Indian Springs Formation. Conodont studies by Dunn (1965, 1966) and, more particularly, by Webster (1969) and by Lane, Sanderson and Verville (1972) clearly indicate that the *Rhipdomella nevadensis* megafaunal zone, characteristic of the Indian Springs Formation and basal Bird Spring Formation, is best considered of Chesterian age.

The lithostrotionid biostromes, well exposed in this section, characterize the uppermost part of the Yellowpine Limestone throughout the Arrow Canyon Range, Las Vegas Range, and Meadow Valley Range. The basal contact of the Yellowpine Limestone is here marked by a slight reentrant in the gorge wall, separating the somewhat less thickly bedded ledges of the Yellowpine Limestone from the very thick, massive beds of Bullion Limestone in the deepest and narrowest parts of the gorge.

0.2

63.8 Prominent petroglyphs on north wall of gorge. Cherty beds of Anchor-type lithology occur within the thick Bullion Limestone cliffs a bit beyond the petroglyphs. In the Arrow Canyon Range sections, which are somewhat more basinward than those in Frenchman Mountain, at Apex, in the Muddy Mountains, and at Mountain Spring Pass, there are several cherty sequences separating substantial thicknesses of Bullion-type lithology. As a consequence, a clear-cut distinction between the Anchor Limestone and Bullion Limestone is not readily made here and it is entirely possible that, eventually, a more complex rock-stratigraphic classification will be required to deal with the Arrow Canyon Range exposures of the Monte Cristo group. Turn and retrace route out of Arrow Canyon to Nevada 7.

1.5

65.3 Good end-on view of Monte Cristo strike ridge crossing southern end of Mormon Mountains at 12 o'clock. The sharply pointed peak is at an elevation of 6,505 feet (2,000 m); Mormon Mountain, the highest peak in the Mormon Mountains, has an elevation of 7,411 feet (2,250 m).

1.3

66.6 Rejoin paved road at mouth of Arrow Canyon; turn left.

0.4

67.0 Junction with Nevada 7, turn left more than 90°, climbing out of Moapa Valley.

0.7

67.7 Muddy Creek Formation excellently exposed at 3 o'clock. Hills on either side of highway ahead are in Bird Spring Formation; as we proceed, we will cross the same structure traversed in Arrow Canyon.

1.9

69.6 Reasonably well-exposed outcrops of Indian Springs Formation at 1 o'clock. The characteristic red soil has attracted the attention of prospectors in this area — to no avail.

0.9

70.5 Road ascends to level of Pleistocene alluvial surface. Well-developed caliche crust and desert varnish. Sheep Range ahead at 12 o'clock.

0.1

70.6 **STOP 4 — NORTH END, ARROW CANYON RANGE.** Fossil collecting on low hill at 2 o'clock is excellent in Anchor Limestone. North end of Arrow Canyon Range, in this view largely displaying dip slopes of Late Devonian Arrow Canyon Formation and Mississippian Monte Cristo Group, visible at 9–10 o'clock. Muddy Creek Formation well exposed in valley along wash between us and Arrow Canyon Range. This is wash that flows into and through Arrow Canyon. Mountains at 1 o'clock are in southern end of Meadow Valley Range and, from this point, expose Ordovician through Devonian rocks.

0.7

71.3 Contact between Late Devonian (Famennian) Crystal Pass Limestone and Early Mississippian (Kinderhookian or Late Tournaisian) Dawn Limestone well exposed at 2 o'clock. Marked contrast between light-gray rocks in Crystal Pass Limestone and black limestone of overlying Dawn Limestone provides a readily recognized stratigraphic datum for the region.

1.8

73.1 Relationships between main ridge of Arrow Canyon Range, eastern foothills ridge, Dry Lake Thrust, and distribution of Paleozoic rocks in range may clearly be seen to south. Dry Lake Thrust (see comments at miles 18.2 and 28.1 earlier today) crosses gap between main range and lower eastern ridges of Arrow Canyon Range. Low ridge extending northwesterly from eastern foothills, in view just below skyline, is on lower plate of Dry Lake Thrust. Late Cambrian Nopah Formation crops out at its northernmost tip; remainder of ridge exposes Early Ordovician Pogonip Group. Remainder of Paleozoic sequence is in stratigraphic continuity within eastern foothills to left. These rocks make up an apparently continuous sequence of miogeosynclinal Paleozoic rocks, ranging in age from Late Cambrian to Middle Permian, that is about 16,000 feet (5,000 m) thick. Synclinal fold exposed in top of flat-topped mountain at western margin of saddle is Arrow Canyon Syncline, here affecting rocks of Late Devonian Arrow Canyon Formation. This is the same flat-topped hill and syncline mentioned in the commentary at mile 28.1 earlier today; you are viewing the mountain from the opposite side. Sequence seen from here in east face of main ridge of Arrow Canyon Range comprises rocks from uppermost Pogonip Group through Late Devonian Arrow Canyon Formation.

Figure 56. **Base of Battleship Wash Formation as exposed on north wall of Arrow Canyon. 1 – abundant lithostrotionid coral heads in uppermost part of Santogrossi's (1977) measured section; 2 – near contact between Santogrossi's (1977) units 17 and 18; 3 – top of Yellowpine Limestone. Rice and Langenheim's (1974) unit 1 of Battleship Wash Formation is in reentrant at base of cliff, but is somewhat lenticular. Remainder of cliff is almost entirely within unit 2 of Rice and Langenheim (1974).**

2.1

75.2 Intersection with old U.S. 93 to right; continue to left on Nevada 7, descending into valley.

0.3

75.5 Northernmost end of Arrow Canyon Range, below us at 12 o'clock, is composed of Mississippian Dawn and Anchor Limestones dipping toward us. To west, northern part of Las Vegas Range interrupts magnificent alluvial fan slope and is composed of Bird Spring Formation in characteristic terraced outcrop. Sheep Range, here composed of Cambrian and Ordovician rocks, forms scarp at head of fan slope and skyline. Gass Peak Thrust crops out along foot of Sheep Range.

0.8

76.3 View to north up Coyote Springs Valley and upper Muddy River or lower White River. The stream is intermittent here and is part of the White River drainage basin, and the river names are not consistently applied. All of this valley drains through Arrow Canyon. Low hills at 10 o'clock are outcrops of Monte Cristo Group and are the northernmost exposures in Arrow Canyon Range, main ridge structural block.

0.4

76.7 Cross wash of Muddy or White River. Newly patched highway here replaces a washed-out segment lost during the last major flash flood.

3.6

80.3 Junction of Nevada 7 and U.S. 93; turn left. Low range at 12 o'clock (orientation given before making turn) is Las Vegas Range, here composed of complexly folded Bird Spring Formation. Prominent, rusty-weathering layer is "Tungsten Gap Chert." Delamar Range forms skyline at 3:30. Flat-lying rocks on crest of Delamar Range are part of an Eocene-Oligocene welded tuff-bearing sequence resting in angular unconformity on Paleozoic rocks. Thick, white-weathering layer in tuff sequence is Hiko Tuff. Acidic to intermediate volcanic rocks of early Tertiary age, characterized by ash-flow tuffs, are widespread throughout the Great Basin and probably covered the entire region. Paleozoic rocks below tuff in Delamar Range comprise a complete miogeosynclinal sequence from within Cambrian through Bird Spring Formation. Broad wash at 4 o'clock is Kane Springs Wash, separating Delamar and Meadow Valley Ranges. Mountains at 4:30-6:00 are on west face of Meadow Valley Range, a continuation of structural block carrying eastern ridges of Arrow Canyon Range and entire Dry Lake Range. Prominent black outcrop is that of Middle and Late Ordovician Ely Springs Dolomite. Silurian and Devonian rocks crop out visibly above, and a bit of Early Ordovician Pogonip Group is seen below. Distant peak on horizon, viewed through gap between Arrow Canyon Range and Meadow Valley Mountains, is Virgin Peak, elevation 8,064 feet (2,450 m). This peak is capped by a Paleozoic platform sequence not unlike that at Frenchman Mountain, but the bulk of the mountain consists of Precambrian rock. West face of main ridge of Arrow Canyon Range occupies 6:30-9:00 quadrant. This fault-line scarp exposes a complete sequence from Late Cambrian Nopah Formation through Mississippian Anchor Limestone in one of the best and most intensively studied Paleozoic sections in the region (Langenheim et al., 1962). High (elev. 4,650 feet (1,415 m)), notched peak at 9:00-9:30 is an eastern outlier of Las Vegas Range. Cliffs at top are mostly of Bullion Limestone and Yellowpine Limestone, with lower slopes exposing remainder of Mississippian Monte Cristo Group, Crystal Pass Limestone, and Arrow Canyon Formation. Deformed rocks of Bird Spring Formation form hills of Las Vegas Range between Monte Cristo Group in outlier peak and 12 o'clock and beyond to north of us for a short distance.

2.0

82.3 Mormon Wells Road to right. Mormon Wells Road crosses Las Vegas Range and Sheep Range to reach headquarters of Desert Game Refuge at Corn Springs on U.S. 95 in Las Vegas Valley. It is a primitive road, barely passable to ordinary vehicles, with great care and luck, but it is easily traveled in a light truck or carryall-type vehicle. There is little traffic and there are no facilities short of Corn Springs. Continue on U.S. 93.

1.2

83.5 **STOP 5—ARROW CANYON SEQUENCE VIEW.** Pull off to view the Arrow Canyon Range sequence from the fan crest (figs. 57, 58).

4.0

87.5 Primitive road to left; continue straight ahead on U.S. 93. High (elev. 5,100 feet (1,550 m)) mountain to left is topped by black cliffs of Ely Springs Dolomite resting on pink Eureka Quartzite. Remainder of mountain face is composed of Pogonip Group resting on black dolomite of Late Cambrian Nopah Formation. Lower half of Pogonip Group is dolomitic, weathers rubbly, and is roughly equivalent to Goodwin Formation. Cliffs and benches above, better exposed at last stop, are composed of limestone that is roughly equivalent to Antelope Valley Limestone. Inconspicuous, low, rounded brownish hill (elev. 2,880+ feet (700 m)) in bottom of valley at 11:30 is composed of Silurian Laketown Dolomite with a thin capping of basal Devonian sandstone in Piute Formation. Provided that this low hill is not an isolated fault block or a landslide mass, its relationship to the mountain front establishes the stratigraphic displacement of the Arrow Canyon Range range-front fault at this point — about 3,300 feet (1,000 m).

2.4

89.9 **STOP 5—LOWER SECTION, ARROW CANYON RANGE.** Pull off to view lower part of Arrow Canyon Range section and range-front fault (fig. 59).

1.1

91.0 Silurian and Devonian rocks in low hills below road at 9 o'clock are here in their closest proximity to outcrops of Arrow Canyon Range in cliffy slopes at 3 o'clock.

3.4

94.4 Quarry in Eureka Quartizite at 9 o'clock. Hills behind and around quarry are in complexly faulted Ordovician to Devonian rocks. Monte Cristo Group underlies hills at 3 o'clock in an outlier of Las Vegas Range.

1.0

95.4 Roadcut in Monte Cristo Group limestone. Hidden Valley Playa at 12 o'clock is in center of a small basin of internal drainage.

1.3

96.7 Light-gray to white-weathering layer at 9 o'clock is Late Devonian Crystal Pass Limestone with black outcrops of Monte Cristo Group above. Brenckle's (1973) Arrow Canyon Range section is located here in ridges to south along east side of playa. Brenckle's section more or less duplicates section reported by Langenheim et al. (1962). Webster's (1969) Dry Lake section is located along upper part of same ridge in which sections of Langenheim et al. (1962) and Brenckle (1973) end, and continues through Indian Springs Formation to top of Atokan in Bird Spring Formation. Others who have reported on parts of this section include Langenheim and Collinson (1963) on Crystal Pass Limestone, Rice and Langenheim (1977) on Battleship Wash Formation, and Santogrossi (1977) on Yellowpine Limestone.

1.4

Figure 57. Panoramic view of part of west face of Arrow Canyon Range. 1 – light-colored outcrop of Middle Ordovician Eureka Quartzite. 2 – top of Middle and Late Ordovician Ely Springs Dolomite, which includes light-colored bench at top of prominent black cliff, as well as black dolomite in cliff. 3 – contact between Silurian Laketown Dolomite and Early (?) and Middle Devonian Piute Formation. Sequence between 2 and 3 is repeated in part as a consequence of faulting subparallel to face of scarp. 4 – base of Late Devonian (?) Moapa Formation; 5 – base of Late Devonian Arrow Canyon Formation; 6 – base of Late Devonian Crystal Pass Limestone; 7 – base of Early Mississippian Dawn Limestone.

Figure 58. Panorama of a part of west face of Arrow Canyon Range. 1-4 – tops of "formations" Op_c through Op_f, respectively, of Langenheim et al. (1962), belonging to Early Ordovician Pogonip Group. 5 – top of Middle Ordovician Eureka Quartzite; 6 – top of Middle and Upper Ordovician Ely Springs Dolomite; 7 – top of Silurian Laketown Dolomite; 8 – top of Early (?) and Middle Devonian Piute Formation; 9 – contact between Late (?) Devonian Piute Formation and Late Devonian Arrow Canyon Formation.

Figure 59. View of west face of Arrow Canyon Range opposite highest peak in range. This scarp and the fan below document nearly all of the stratigraphic displacement on the fault bounding the range. 1 – top of Late Cambrian Nopah Formation; 2 – top of Op_a "formation" of Langenheim et al. (1962). 3 – top of cliff-forming Op_c "formation" of the Pogonip Group. Op_c "formation" rests on a bench-forming unit, the Op_b "formation," and is succeeded by another bench-forming unit, the Op_d "formation." All of the Pogonip Group is assigned to Early Ordovician. 4 – contact between Op_e "formation," a cliff former, and uppermost Op_f "formation" of Pogonip Group. 5 – top of a sub-summit of Eureka Quartzite in a fault slice fronting scarp. 6 – peak, almost at top of Middle and Late Ordovician Ely Springs Dolomite.

98.1 Road now on playa surface at 2,649 feet (810 m) elevation.

2.1

100.2 Bird Spring Formation, chiefly the Pennsylvanian portion, crops out on slopes to left in complexly faulted exposures. Rusty-weathering "Tungsten Gap Chert" unit is conspicuous and has been a "key bed" in mapping here. All exposures in Las Vegas Range to west, as seen from here, are in Bird Spring Formation. Apparent continuity of sequence, however, is deceptive and close examination should reveal apparent convergence of strata above and below axes of superposed "chevron" folds striking across face of range.

3.4

103.6 Muddy Peak and Buffington Pockets are visible on horizon at 10 o'clock beyond Arrow Canyon Range. Las Vegas and Sheep Ranges are visible to right.

1.6

105.2 Road turns left. Note Arrolime quarries to right of road and well ahead. U.S. 93 continues across Dry Lake Valley to rejoin I–15. Dry Lake Range and Muddy Peak are ahead with Muddy Peak the highest point on skyline.

1.7

106.9 Virgin Peak visible at 10 o'clock as a high point on skyline. This mountain consists of a platform sequence of Paleozoic rocks resting on a Precambrian base.

4.3

111.2 Turn right on interchange to proceed southwesterly to Las Vegas on I–15. Continue through underpass and turn left to take I–15 to Utah.

END OF SEVENTH DAY ROAD LOG

STOP DESCRIPTIONS — SEVENTH DAY: CLARK COUNTY, NEVADA

Gary D. Webster[1] and R. L. Langenheim, Jr.[2]

STOP 1—DRY LAKE VALLEY
NW¼, sec. 13, T. 18 S., R. 64 E.
Clark County, Nevada

We are in Dry Lake Valley between the Dry Lake Range and the Arrow Canyon Range and for the next few miles will be traveling essentially along the trace of the Dry Lake Thrust. The Dry Lake Thrust is the second thrust fault cropping out northwest of the Muddy Peak Thrust as mapped by Longwell et al. (1965). The thrust more or less defines the west side of the Dry Lake Range and passes between the main ridge of the Arrow Canyon Range and its foothills to the east. Projection of its trace to the north suggests that it continues as the Delamar Mountain Thrust and Fold Belt of Tschanz and Pampeyan (1970) after passing beneath alluvium in White River Valley and Kane Springs Wash north of the Arrow Canyon Range.

The Dry Lake Thrust is exposed at two localities: at Apex Summit and in the Arrow Canyon Range. At Apex, it brings Pogonip Group rocks in the overthrust plate into contact with Bird Spring Group in the lower plate. In the Arrow Canyon Range, the Nopah Formation in the upthrust block rests on inverted Pogonip Group in the downthrown block. At this point the fault plane dips about 60° northwest and the fault is actually a high angle reverse fault.

The Glendale Thrust of Longwell et al. (1965) is mapped along the northwest face of the Muddy Mountains, which are not visible from this point, and the sole fault, the Muddy Peak Thrust, is exposed in the southern part of the Muddy Mountains (figs. 23, 28). The latter, Muddy Peak Thrust, is generally considered equivalent to the Keystone Thrust seen yesterday. Consequently, at Stop 1 we are farther northwest into the thrust belt and into the miogeosyncline than we reached in yesterday's trip.

Structural features of the thrust belt northwest of Stop 1 and to be observed today include the Las Vegas disturbed belt and the Gass Peak Thrust. The Las Vegas disturbed belt is a series of "chevron" type folds cropping out in the western part of the Las Vegas Range. These folds, mapped as part of the Gass Peak Thrust and Fold Belt by Tschanz and Pampeyan (1970) in Lincoln County, are traceable from the south end of the Las Vegas Range northward into southern Lincoln County. The Gass Peak Thrust, which is a few miles west of the fold belt, also is a high-angle reverse fault where exposed along the base of the Sheep Range.

Most of the terrain visible from Stop 1 in the Arrow Canyon and Dry Lake Ranges is Carboniferous. The massive cliffs and dark-colored rocks across Dry Lake Playa and ahead of us in the Arrow Canyon Range are Monte Cristo Group rocks resting on Arrow Canyon Formation and Crystal Pass Limestone. The markedly banded outcrops on all sides of us are Bird Spring Formation. The entire pre-Devonian and post-late Cambrian regional sequence, however, is exposed in the rocks above and below the trace of the Dry Lake Thrust ahead of us in the eastern foothills of the Arrow Canyon Range.

A well drilled in the center of Dry Lake Playa penetrated 310 feet (94 m) of Pleistocene and recent playa sediments and 1,190 feet (360 m) of Muddy Creek Formation without entering pre-Pliocene rocks. A second well, about three-fourths of a mile in front of the mountains on the west side of the playa penetrated 35 feet (10 m) of recent alluvium, 150 feet (45 m) of Pleistocene playa deposits, and 773 feet (235 m) of Pleistocene Muddy Creek Formation. The Muddy Creek Formation in these wells contains substantial amounts of gypsum interbedded with clay and bentonitic clay.

STOP 2 — GLENDALE THRUST
SE¼, sec. 16, T. 15 S., R. 66 E.
Clark County, Nevada

Since leaving Stop 1, we have crossed the trace of the Dry Lake Thrust and are now on the block between the Dry Lake Thrust and the Glendale Thrust. The Glendale Thrust is exposed in only a few scattered localities: on the west face of the Muddy Mountains just north of the Buffington Pockets, at the north end of the North Muddy Mountains, and in the hills north of Glendale between the North Muddy Mountains and the Mormon Mountains (Longwell et al., 1965, pl. 5).

STOP 3 — ARROW CANYON
E½, sec. 11 & S½, sec. 12, T. 14 S., R. 64 E. & SW¼, sec. 7, T. 14 S., R. 65 E.
Clark County, Nevada

Stop 3 includes exposure of the Monte Cristo Group, Battleship Wash Formation, Indian Springs Formation, and Bird Spring Formation as exposed in Arrow Canyon.

The lower part of the cliff-forming, massive, black limestone sequence comprising the Monte Cristo Group and Battleship Wash Formation is complicated by folding and faulting in the upper reaches of Arrow Canyon and will not be visited during our trip. We will examine only the uppermost Bullion Formation, Battleship Wash Formation, Indian Springs Formation, and the Mississippian and Pennsylvanian parts of the Bird Spring Formation.

Stratigraphically continuous, well-exposed sections of the Monte Cristo Group, however, are exposed at numerous points between Battleship Wash, the first east-west trending valley south of Arrow Canyon, and the southern end of the ridge of overturned sediments below the Dry Lake Thrust, as viewed from Stops 1 and 2. The section at "Tungsten Gap," located in the N½, sec. 10, T. 64 E., R. 15 S. (approx.), and about 5½ miles south of Arrow Canyon, has been thoroughly described from well below the base of the Monte Cristo Group through the top of the Battleship Wash Formation (fig. 60) (Hansen and Carozzi, 1974; Castle, 1967; Langenheim et al., 1962; Pierce, 1969; Pierce and Langenheim, 1972; Rice and Langenheim, 1974; Ryan and Langenheim, 1973; Waines, 1965).

[1] Department of Geology, Washington State University, Pullman, Washington 99163.
[2] Department of Geology, University of Illinois, Urbana, Illinois 61801.

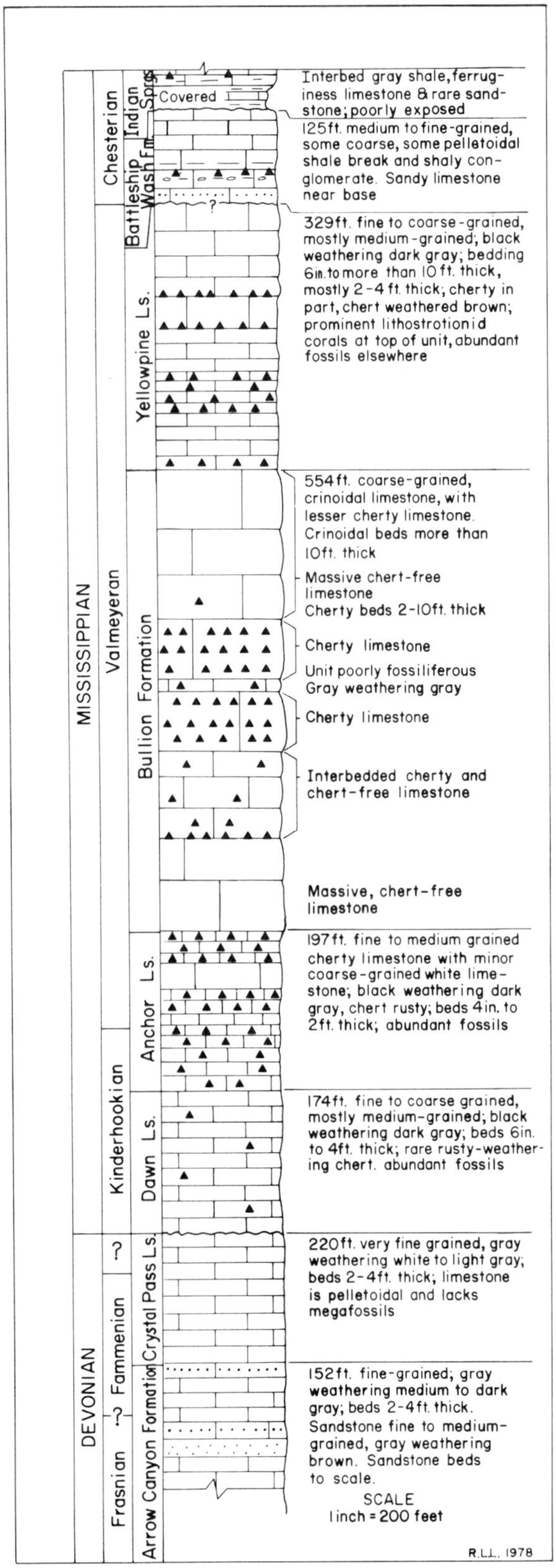

Figure 60. Columnar section of Crystal Pass Limestone through Battleship Wash Formation at Tungsten Gap. Compiled from work of Castle (1967), Langenheim et al. (1962), Pierce (1969), Rice and Langenheim (1974), Ryan and Langenheim (1973), and Waines (1965).

As in the Valentine Limestone Member of the Sultan Limestone, which we saw yesterday at Mountain Springs Pass, the upper member of the Arrow Canyon Formation includes several brown-weathering sandstone beds within a sequence of very fine- to fine-grained, mostly pelletoidal limestone that is gray and weathers medium to dark gray. The contact with the Crystal Pass Limestone above is gradational in detail but appears abrupt when viewed from a distance. The contact is taken at the point at which the continuous sequence of light-gray to white-weathering pelletoidal dolomite begins, a bedding plane somewhere not too many feet above the highest persistent sandstone in the uppermost Arrow Canyon Formation (Ryan and Langenheim, 1973). Extrapolation of conodont occurrences described by Reinbold and Langenheim (1977) in the Las Vegas Range at around mile 85 on today's log indicates that the upper member of the Arrow Canyon Formation is of Frasnian to Famennian age. The overlying Crystal Pass Limestone, 220 feet (67 m) thick at this locality, is also mostly, if not entirely, of Famennian age. Conodonts studied in the Las Vegas Range by Reinbold and Langenheim (1977) and in the Arrow Canyon Range by Langenheim and Collison (1963) indicate that the entire unit, with the exception of the uppermost 50 feet (15 m) is of Famennian age. Brenckle (1973), however, on the basis of occurrences of endothyroid foraminifers assigns the uppermost 55 feet (17 m) of the Crystal Pass Limestone at a locality near that of Langenheim and Collinson (1963) to the Kinderhookian. Mamet and Skipp (1970) reported the upper part of the Crystal Pass Limestone to include Mamet foraminiferal zone 7, considering it of early Osage age. Inasmuch as Brenckle's age assignment is based on a single occurrence of probable late Kinderhookian foraminifers, Mamet and Skipp's data is a preliminary report, and because there are no conodonts in that uppermost fifty feet or so of Crystal Pass Limestone, the paleontologic data appear meager. A lens of sandstone and sandy dolomite does occur within the Crystal Pass Limestone at about 50 feet (15 m) below the upper contact with the Dawn Limestone (Reinbold and Langenheim, 1977) and may be evidence for an unconformity at that point, separating Kinderhookian from not quite uppermost Famennian rocks. On the other hand, there is clear-cut evidence for an erosion surface separating the top of the Crystal Pass Limestone from the overlying Dawn Limestone, and this may be the disconformity separating Devonian from Mississippian rocks in the Arrow Canyon section.

The overlying Dawn and Anchor Limestones are not strikingly different from their aspect at Mountain Springs Pass. The Dawn Limestone is about 100 feet (30 m) thicker as recognized at Tungsten Gap, but the Anchor Limestone is about 25 feet (8 m) thinner than at Mountain Springs Pass. Thus, there is no great difference in thickness or in lithologic character between these units in passing from the platform edge at Mountain Springs into the miogeosyncline in the Arrow Canyon Range. Mamet and Skipp (1970) recognized foraminifers of upper zone 8 in the Dawn and foraminifers of zone 9 and the lower two-thirds of zone 10 in the Anchor, placing the Osage-Meramec boundary within the Anchor Limestone. Brenckle (1973) placed the Dawn Limestone in the Osage, recognizing within it foraminifers of the upper half of Mamet zone 7 and 8. He also considered the Osage-Meramec boundary to be in the Anchor Limestone and reported foraminifers of Mamet zone 9 in the lower part of the Anchor and undiagnostic foraminifers in the middle and upper Anchor Limestone (Brenckle, 1973).

Pierce and Langenheim (1972) recognize *Polygnathus communis communis* and *Pseudopolygnathus dentilineatus* ranging throughout the entire Dawn Limestone and the lower part of the Anchor Limestone, assigning these rocks to the zone of that name (fig. 61). In addition, *Pseudopolygnathus primus* occurs in the uppermost 68 feet assigned to the zone in the Anchor Limestone.

The middle 25 feet (8 m) of the Anchor Limestone are assigned to the Osagian or early Valmeyeran *Polygnathus communis carinus* Zone on the occurrence of *P. communis communis, P. communis carinus, P. dentilineatus* and *P. primus*, which occur together below the lowest occurrence of *Gnathodus semiglaber.* This fossil defines the base of the *Gnathodus semiglaber-Pseudopolygnathus marginatus* Zone which comprises the remainder of the Anchor Limestone and the lowermost Bullion Limestone.

The Bullion Limestone is almost 200 feet (60 m) thicker in the geosynclinal Arrow Canyon Range sections than in the platform margin section at Mountain Springs Pass. In addition, substantial intervals of cherty limestone, having the same general character as the Anchor Limestone, occur within the unit as defined here by Langenheim et al. (1962).

Based on foraminifers, Mamet and Skipp (1970) considered the Bullion Limestone to be Meramecian containing foraminiferal zones from upper 10 through 13. Brenckle (1973) reported questionable zone 10 foraminifers in the middle of the Bullion Limestone and the upper limit to be zone 12.

Gnathodus cuneiformis, G. typicus and *Pseudopolygnathus marginatus* also appear near the base of this zone along with *Pseudopolygnathus triangulis triangulis* and *Gnathodus bilineatus* which, however, become more abundant toward the top of the zone. The *Gnathodus bilineatus-Gnathodus cuneiformis* Zone, the next highest zone in the Bullion Limestone, is marked at the base by the lowest abundant occurrence of *G. bilineatus* and the highest abundant occurrence of *G. cuneiformis*. A new subspecies of *Pseudopolygnathus triangulis* n. subsp. a., appears in abundance in the upper third of the zone and continues beyond the highest occurrence of *Gnathodus bilineatus. P. triangulis* n. subsp. a. is abundant to common in the lower portion of the overlying zone of *P. triangulus* n. subsp. a. where it is accompanied by *P. triangulis triangulis, Polygnathus communis communis* and *Pseudopolygnathus marginatus* which continue abundant throughout the zone. The zone of *Polygnathus triangulis* n. subsp. a. comprises the middle half of the Bullion Limestone. The lowest occurrence of *Eotaphrus burlingtonensis* marks the base of the next highest zone, the *Polygnathus communis communis-Eotaphrus burlingtonensis* Zone, comprising the lower half of the upper part of the Bullion Limestone. Above this zone, the remainder of the range of *Eotaphrus burlingtonensis*, accompanied by the earliest occurrence of *Taphrognathus varians* defines the *Eotaphrus burlingtonensis* Zone. This zone includes the uppermost Bullion Limestone and lowermost Yellowpine Limestone.

The Arrowhead Limestone is absent in this geosynclinal sequence, but the Yellowpine Limestone is much thicker than at Mountain Springs Pass. In addition, the Yellowpine Limestone in the Arrow Canyon Range is more cherty and is characterized by a prominent lithostrotionid coral biostrome at its top (Santogrossi, 1977). Corals occurring here include *Lithostrotion (L.) packardi, L. (Siphonodendron) warreni, L. (S.)* cf. *L. (S.) whitneyi, Lithostrotionella banffensis, L. mclareni, L. microstylum* and *L. shimeri*; along with *Ekvasophyllum turbineum* and *Turbophyllum multiconem* near the top of the biostrome.

Foraminifers of zone 14 and the lower half of zone 15 were reported from the Yellowpine Limestone by Mamet and Skipp (1970). Brenckle (1973) reported zone 13 foraminifers in the base of the Yellowpine Limestone, suggested that zone 14 faunas might be present, and recognized rich foraminiferal faunas of the lower part of zone 15 in the upper third of the Yellowpine Limestone.

The middle half of the Yellowpine Limestone at Tungsten Gap is in the *Taphrognathus varians - Polygnathus mehli* Zone, the base of which is placed above the highest occurrence of *Eotaphrus burlingtonensis.* The uppermost part of the Yellowpine Limestone, including the coral biostromes, is in the *Taphrognathus varians* Zone, which is probably early Meramecian. In addition to *T. varians, Cavusgnathus* occurs sparingly and there are no *Polygnathus* spp.

The Battleship Wash Formation, of course, is not present in the platform margin section at Mountain Springs Pass, but is 125 feet (38 m) thick at Tungsten Gap. In the lower half, the *Taphrognathus varians - Apatognathus* and *Gnathodus bilineatus* Zones as recognized by Rice and Langenheim (1974) complete the Meramecian sequence. These rocks also contain *Faberophyllum* spp. (Duley, 1957; Rice and Langenheim, 1974) belonging to the "F" zone of Sando, Mamet and Dutro

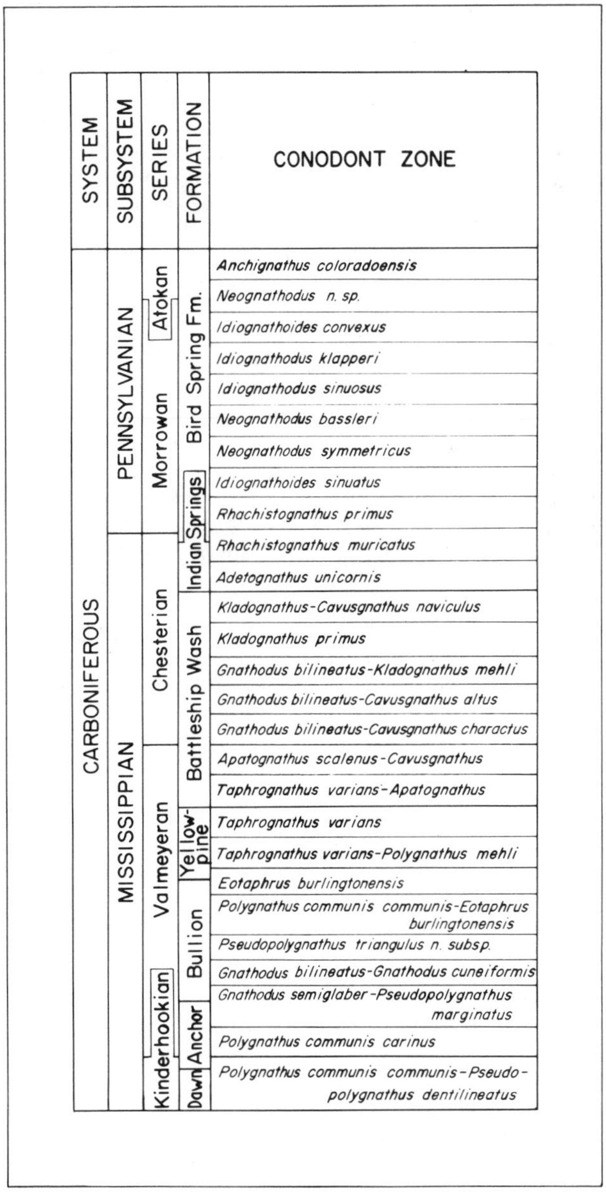

Figure 61. Carboniferous conodont zones recognized in southern part of Cordilleran Miogeosyncline. Data compiled after Webster (1969), Dunn (1970), Lane et al. (1971), Pierce and Langenheim (1972), Rice and Langenheim (1974), Lane and Straka (1974), Dunn (1976), and Lane (1977).

(1969). The uppermost part of the Battleship Wash Formation, however, is Chesterian, belonging to conodont zones in uninterrupted sequence from the *Gnathodus bilineatus-Cavusgnathus charactus* Zone through the *Klodognathus-Cavusgnathus naviculus* Zone. A few horn corals and the *Stigmaria wedingtonensis* locality of Arrow Canyon occur at the top of the formation. Foraminifers of the upper half of zone 15, lower part of zone 16i, and zone 17 were reported from the Battleship Wash Formation by Mamet and Skipp (1970). Brenckle (1973), however, did not recognize any zone 16i forams, but reported a barren 40 foot (12 m) interval between his zones 15 and 17-18. Thus, both conodonts and foraminifers suggest an unconformity within the early Chesterian part of the Battleship Wash Formation.

The overlying Indian Springs Formation is nearly four times thicker at Arrow Canyon (220 feet (67 m)) than Mountain Springs Pass (59 feet (18 m)). Although bedding is thin and many lithologic types are present in the Indian Springs Formation, it is dominantly dark-gray shale in the lower three-fourths and ferruginous limestone in the upper one-fourth. Only the marker conglomerate (seen yesterday at Stop 4, Apex Section), near the middle of the formation, may be traced throughout the Arrow Canyon Range and into some nearby mountain ranges (Webster, 1969). The Indian Springs Formation yields a diverse megafauna, especially rich in brachiopods, with moderately abundant corals, bryozoans, gastropods, bivalves, blastoids, and crinoids. Conodonts, foraminifers and ostracodes are abundant in the microfauna.

The Indian Springs Formation is in the lower part of the *Rhipidomella nevadensis* Assemblage Zone (Webster, 1969) and Mamet foraminiferal zones 18 and 19 (Mamet and Skipp, 1970) or only zone 19 (Brenckle, 1973). Also, the Indian Springs Formation includes the *Adetognathus unicornis* Zone, as restricted by Lane and Straka (1974), and the lower half of the *Rhachistognathus muricatus* Zone, as originally defined by Lane and Straka (1974). Webster (1969) proposed the *Gnathodus girtyi simplex* Zone, which included the Indian Springs Formation and Chesterian part of the Bird Spring Formation. Recognition of the *A. unicornis* and *R. muricatus* Zones for the same chronologic section allows greater definition, therefore the *G. girtyi simplex* Zone as defined by Webster (1969) should be abandoned, as suggested by Lane and Straka (1974). Dunn (1970) restricted the *G. girtyi simplex* Zone to the uppermost part of the *Adetognathus unicornis* Zone reporting *G. girtyi simplex* replacing *G. girtyi girtyi* in the Indian Springs Formation. *G. girtyi simplex,* distingushed by a single node in the short outer parapet, first occurs in the uppermost part of the Battleship Wash Formation with *G. girtyi girtyi.* Basing the *G. girtyi simplex* Zone on first occurrence would make it synonomous with the *A. unicornis* Zone; therefore, the restricted *G. girtyi simplex* Zone of Dunn (1970) is not accepted.

Approximately 2,700 feet (820 m) of the Bird Spring Formation is in nearly continuous exposure in Arrow Canyon. This includes almost all of the BSc and all of the BSd and BSe units of Langenheim and Langenheim (1965) and Cassity and Langenheim (1966). Strata range in age from late Chesterian into Wolfcampian and the top is removed by erosion. In maximum development, such as at Lee Canyon on the north side of the Spring Mountains, the Bird Spring Formation is approximately 7,000 feet (2,130 m) thick and continues into the Leonardian (Rich, 1961). In Arrow Canyon the Bird Spring Formation is thicker than equivalent sections observed at Stops 2, 4, and 6 yesterday.

The Bird Spring Formation is dominantly biomicrite and biosparite (many with arenaceous or argillaceous content), with thin shale beds and sandstone. Nodules and discontinuous bands of chert are common in many beds. Dolomite is common in the Permian part of the section. Except for the "marker limestone" of Webster (1969) and the "Tungsten Gap Chert" of Castle (1967) and of Micklin (1967), individual lithologic units are not laterally continuous in the Bird Spring. Thus, most stratigraphic sections only a few miles apart are correlated by fossils.

The basal 48 feet (15 m) of the Bird Spring Formation are latest Chesterian in age, included in the *Rhipidomella nevadensis* Assemblage Zone by Webster (1969), the *Rhachistognathus muricatus* Zone by Lane and Straka (1974), and Mamet foraminiferal zone 19 (Mamet and Skipp, 1970; Brenckle, 1973). A thin basal conglomerate marks the Mississippian-Pennsylvanian boundary (Webster and Lane, 1967; Webster, 1969).

Differences of opinion exist concerning the thickness and development of Morrowan strata in the Bird Spring Formation. Mamet and Skipp (1970, fig. 3) show an unconformity with only the upper half of zone 20 represented. Although this may be true in some areas of southern Nevada, such as Indian Springs (Webster and Lane, 1967; Gordon and Poole, 1969), it is not believed typical of the Bird Spring Formation and definitely incorrect at Arrow Canyon. Conodonts of the *Rhachistognathus primus* Zone immediately above the basal conglomerate suggest basal Morrowan with little, if any, time gap above the Chesterian (Webster, 1969; Dunn, 1970; Lane et al., 1972; Lane and Straka, 1974).

Three conodont zonations have been proposed for Morrowan strata of the Bird Springs. Webster (1969) proposed the *Streptognathodus noduliferus - Idiognathoides convexus* Assemblage Zone for the lower and middle Morrowan section as he recognized it. This zone should be abandoned because more refined zones are now recognized for the same chronologic section. Dunn (1970) proposed 5 zones including Morrowan strata as he recognized it, and Lane et al. (1972), modified by Lane and Straka (1974), introduced 7 zones as they recognized the Morrowan. Lane (1977), modifying the zonation of Lane and Straka (1974), included an 8th zone extending into the Atokan or Derryan. Although Lane's (1977) zonation is followed on Figure 61, reservations exist about this zonation, including those expressed by Dunn (1976). Neither *Neognathodus bassleri bassleri* nor *Idiognathoidus klapperi* have been found in southern Nevada (reported specimens of *N. bassleri* by Webster, 1969, are *N. roundyi* or *N. colombiensis* and *N.* n. sp.). Strata chronologically equivalent to the *N. bassleri bassleri* and *I. klapperi* Zones are believed present at Arrow Canyon. Before complete acceptance of a conodont zonation for the Morrowan, identifications and ranges given by various workers of Pennsylvanian conodonts must be verified.

Upper limits of Morrowan strata at Arrow Canyon were based on the first occurrence of *Profusulinella* by Brill (1963), Coogan (1964), Langenheim et al. (1962), Langenheim and Langenheim (1965), Cassity and Langenheim (1966), Webster (1969), and Rich (1971). Manet and Skipp (1970) and Lane et al. (1972) consider the boundary to coincide with the first occurrence of foraminifers of zone 21. Dunn (1976) questions the use of first occurrence of foraminifers of zone 21 for definition of the boundary at Arrow Canyon and places it somewhere between 60 to 90 m above that of Lane, et al. (1972). This corresponds to the base of unit 56 of Webster (1969) which he considered middle Morrowan. Until the Morrowan-Atokan boundary is established in the type sections it is impossible to determine its position in other areas.

Undisputed Atokan or Derryan strata of the *Profusulinella* Zone, Mamet foraminiferal zone 22, and the overlying *Fusulinella* Zone at Arrow Canyon (Fig. 62) have been reported by several workers as noted above. However, the only published systematic study of *Profusulinella* and younger fusulinids at

Arrow Canyon is that of Cassity and Langenheim (1966). Their study clearly showed that the zones of *Profusulinella* through *Pseudoschwagerina* as defined by Thompson (1948) are well developed at Arrow Canyon and abundant fusulinids occur at numerous horizons in the Atokan through Wolfcampian parts of the Bird Spring (Cassity and Langenheim, 1966). Detailed stratigraphic columns showing important fusuline and other fossil occurrences are given in Figures 62–64.

Conodonts are abundant in the *Profusulinella* through *Triticites* Zones at Arrow Canyon (Webster, 1968, 1969). Although a zonation has not been proposed for the post-Morrowan section in the Great Basin, species of common Pennsylvanian platform genera such as *Idiognathodus, Idiognathoides, Neognathodus, Adetognathus* and *Streptognathodus* can be used for zonation. The zonation proposed by Merrill (in Lane et al., 1971) does not appear to be applicable in the Great Basin. This is probably due to environmental control of the species present in the two areas.

STOP 4—NORTH END, ARROW CANYON RANGE
NW¼, sec. 35, T. 13 S., R. 64 E.
Clark County, Nevada

Exposures of the Anchor Limestone on the hill north of the highway and along strike for a couple of miles to the north yield abundant fossils. Many specimens are silicified and are easily removed from the matrix by acid-etching techniques. Fossils commonly found at this locality include *Zaphrentites alliptícus, Zaphrentites* sp., *Michelinia expansa, Cyathaxonia*? sp., "*Spirifer*" cf. "*S.*" *centronatus, Rhipidomella*? sp., *Platycrinites* sp. (columnals), *Steganocrinus* cf. *S. pentagonus, Nunnacrinus* sp., and *Agaricocrinus* sp. Acetic acid residues yield specimens of *Polygnathus communis.* The fauna is of Osagian age and correlates to the upper part of the Lodgepole Formation (C_1 zone of Sando et al., 1969) of the Montana shelf and the upper part of the Thunder Springs Member or lower part of the Mooney Falls Member of the Redwall Limestone of northern Arizona.

STOP 5 — ARROW CANYON SEQUENCE VIEW
NW¼, sec. 32, T. 13 S., R. 63 E., unsurveyed
Clark County, Nevada

Silurian through Devonian rocks are exposed at 9 o'clock, with a bit of Monte Cristo Group capping the second peak north of the prominent notch. The Pogonip Group through the Silurian Laketown Dolomite, along with some Devonian rocks, form the scarp at 10 o'clock. The entire Ordovician System rests on Late Cambrian Nopah Formation in the high peak at 11 o'clock. Deformed rocks of the Bird Spring Formation underlie the slopes of the Las Vegas Range to the west, but the low, outlying hill at 12:30 is composed of Monte Cristo Group.

STOP 6 — LOWER SECTION, ARROW CANYON RANGE
SE¼, sec. 28, T. 14 S., R. 63 E.
Clark County, Nevada

For stratigraphy, see Figure 59.

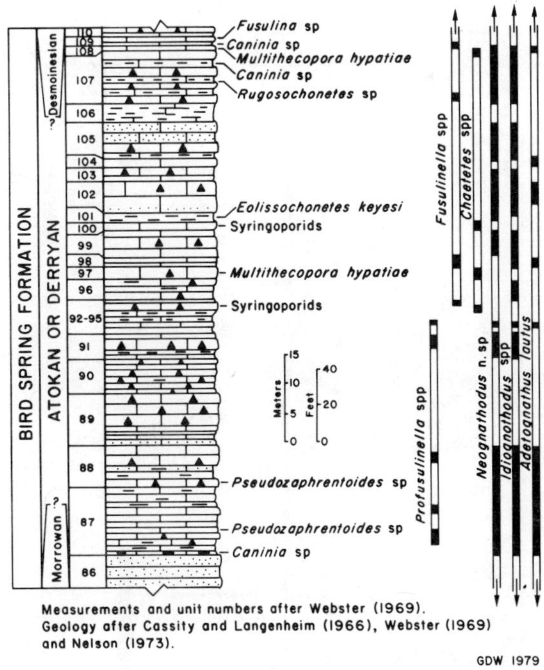

Figure 62. Columnar section of Atokan or Derryan *Profusulinella* and *Fusulinella* Zone rocks exposed in Arrow Canyon. Cassity and Langenheim (1966) place the base of the Desmoinesian approximately 70 feet (21 m) higher in the stratigraphic section.

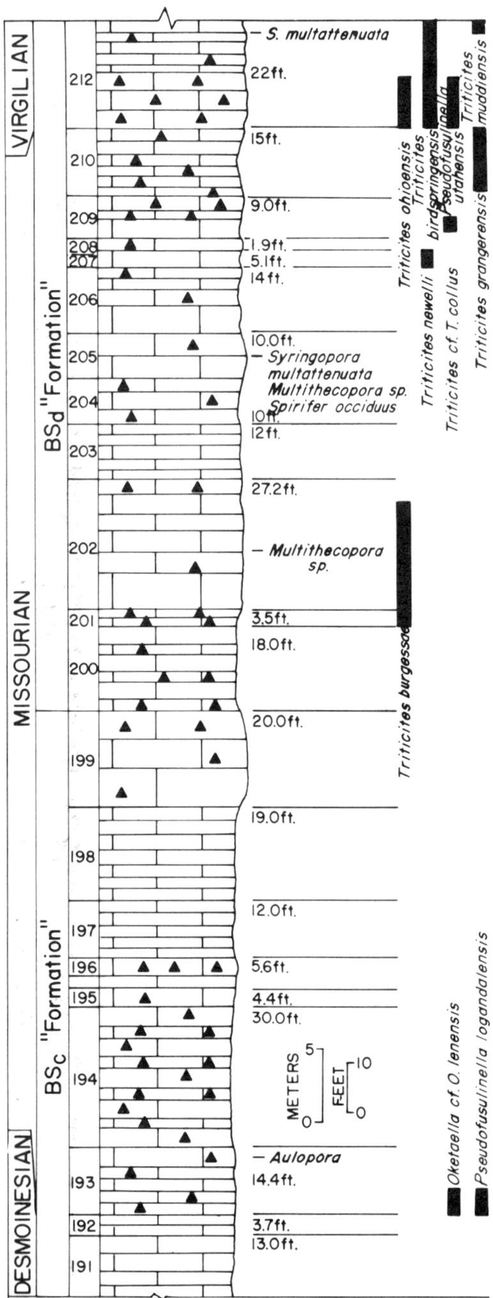

Figure 63. Columnar section of Missourian rocks exposed on north wall of Arrow Canyon.

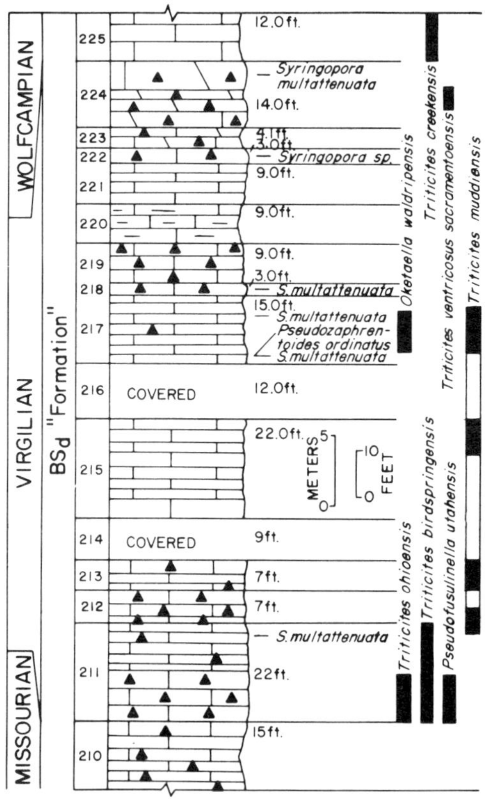

Figure 64. Columnar section of Virgilian rocks exposed on the north wall of Arrow Canyon.

REFERENCES FOR ROAD LOGS AND STOP DESCRIPTIONS

Anderson, R. E., 1969, Notes on the geology and paleohydrology of the Boulder City Pluton, southern Nevada: U.S. Geological Survey Professional Paper 650–B, p. B35–B40.

———1971, Thin skin distension in Tertiary rocks of southeastern Nevada; Geological Society of America Bulletin, v. 82, p. 43–58.

———1973, Large-magnitude Late Tertiary strike-slip faulting north Lake Mead, Nevada: U.S. Geological Survey Professional Paper 794, 18 p.

———et al., 1972 , Significance of K-Ar ages of Tertiary rocks from the Lake Mead region, Nevada-Arizona: Geological Society of America Bulletin, v. 83, p. 273–288.

Barnes, C. W., Eastwood, R., and Beus, S. S., 1974, Geologic resume and field guide north-central Arizona. *In* Karlstrom, N. V., Swann, G. A., and Eastwood, R. L., (eds.), Geology of Northern Arizona: Part II, p. 423–441, Geological Society of America, Rocky Mountain Section Meeting, Flagstaff, Arizona.

Beus, S. S., 1978, Late Devonian (Frasnian) invertebrate fossils from the Jerome Member of the Martin Formation, Verde Valley, Arizona: Journal of Paleontology, v. 52, p. 40–54.

Bissell, H. J., 1962, Permian rocks of parts of Nevada, Utah and Idaho: Geological Society of America Bulletin, v. 73, p. 1083–1110.

Blair, W. N., 1978, Location of the Gulf of California in late Miocene time (abstract): Geological Society of America Abstracts with Programs, v. 10, no. 3, p. 96.

Blair, W. N., McKee, E. D., and Armstrong, A. K., 1977, Age and environment of deposition: Hualapai Limestone Member of the Muddy Creek Formation (abstract): Geological Society of America Abstracts with Programs, v. 9, no. 4, p. 390–391.

Brenckle, P. L., 1973, Smaller Mississippian and Lower Pennsylvanian calcareous foraminifiers from Nevada: Cushman Foundation for Foraminiferal Research Special Publication No. 11, 82 p.

Brill, K. G., 1963, Permo–Pennsylvanian stratigraphy of western Colorado Plateau and eastern Great Basin regions: Geological Society of America Bulletin, v. 74, p. 307–330.

Burchfiel, B. C., et al., 1974 , Geology of Spring Mountains, Nevada: Geological Society of America Bulletin, v. 85, p. 1013–1022.

Carss, B. W., and Carossi, A. V., 1964, Petrology of Upper Devonian pelletoidal limestones: Arrow Canyon Range, Clark County, Nevada: Sedimentology, v. 4, p. 197–224.

Cassity, P. E., and Langenheim, R. L., Jr., 1966, Pennsylvanian and Permian fusulinids of the Bird Spring Group from Arrow Canyon, Clark County, Nevada: Journal of Paleontology, v. 40, p. 931–968.

Castle, R. A., 1967, Mississippian and Pennsylvanian paleontology and stratigraphy at Tungsten Gap North, Arrow Canyon Range, Clark County, Nevada: University of Illinois unpublished B.S. thesis, 56 p.

Coogan, A. H., 1962, Early Pennsylvanian stratigraphy, biostratigraphy, and sedimentation of the Ely Basin, Nevada: University of Illinois unpublished Ph.D. thesis, 90 p.

———1964, Early Pennsylvanian history of Ely Basin, Nevada: American Association of Petroleum Geologists Bulletin, v. 48, p. 487–495.

Damon, P. E., Laughlin, A. W. and Percious, J. K., 1967, Problems of excess argon-40 in volcanic rocks. *In* Radioactive Dating and Methods of Low-Level Counting: Proceedings of International Atomic Energy Agency Symposium, Monaco, p. 463–481.

Duley, D. E., 1957, Mississippian stratigraphy of the Meadow Valley and Arrow Canyon Ranges, southeastern Nevada: University of California at Berkeley unpublished M.A. thesis, 99 p.

Dunn, D. L., 1965, Late Mississippian conodonts from the Bird Spring Formation in Nevada: Journal of Paleontology, v. 30, p. 1145–1150.

———1966, New Pennsylvanian platform conodonts from southwestern United States: Journal of Paleontology, v. 40, p. 1294–1303.

———1970, Conodont zonation near the Mississippian-Pennsylvanian boundary in western United States: Geological Society of America Bulletin, v. 81, p. 2959–2974.

———1976, Biostratigraphic problems of Morrowan and Derryan (Atokan) strata in the Pennsylvanian System of western United States: Geological Society of America Bulletin, v. 87, p. 641–645.

Gans, W. T., 1974, Correlation and redefinition of the Goodsprings Dolomite, southern Nevada and eastern California: Geological Society of America Bulletin, v. 85, p. 189–200.

Gardner, L. R., 1972a, Origin of the Mormon Mesa caliche, Clark County, Nevada: Geological Society of America Bulletin, v. 83, p. 143–156.

———1972b, Pediments and terraces along the Moapa Valley, Clark County, Nevada: Geological Society of America Bulletin, v. 83, p. 3479–3486.

Gordon, Mackenzie, Jr., and Poole, F. G., 1969, Mississippian–Pennsylvanian boundary in southwestern Nevada and southeastern California: Geological Society of America Memoir 110, p. 157–168.

Grove, K. A., 1973, Paleoecologic interpretations of Virgilian fusulinids, Bird Spring Group, Arrow Canyon, Clark County, Nevada: University of Illinois unpublished B.S. thesis, 23 p.

Hansen, M. W., and Carozzi, A. V., 1974, Carbonate microfacies of the Monte Cristo Group (Mississippian), Arrow Canyon Range, Clark County, Nevada: Wyoming Geological Association Earth Science Bulletin, v. 7, p. 13–54.

Heath, C. P., Lumsden, D. N., and Carozzi, A. V., 1967, Petrography of a carbonate transgressive-regressive sequence: The Bird Spring Group (Pennsylvanian), Arrow Canyon Range, Clark County, Nevada: Journal of Sedimentary Petrology, v. 37, p. 377–400.

Hettlinger, C., 1975, Paleoenvironment of syringoporoids (Missourian and Virgilian) of Arrow Canyon, Clark County, Nevada: University of Illinois unpublished B.S. thesis, 39 p.

Hewett, D. F., 1931, Geology and ore deposits of the Goodsprings Quadrangle, Nevada: U.S. Geology Survey Professional Paper 162, 172 p.

Kleinsorge, P. L., 1941, The Boulder Canyon Project: Stanford University Press, 330 p.

Lane, H. R., 1977, Morrowan (Early Pennsylvanian) conodonts of northwestern Arkansas and northeastern Oklahoma: Oklahoma Geology Survey Guidebook 18, p. 177–180.

———and Straka, J. J., II, 1974, Late Mississippian and Early Pennsylvanian conodonts, Arkansas and Oklahoma: Geological Society of America Special Paper 152, 144 p.

Lane, H. R., Merrill, G. K., Straka, J. J., II, and Webster, G. D., 1971, North American Pennsylvanian conodont biostratigraphy. *In* Symposium on conodont biostratigraphy: Geological Society of America Memoir 127, p. 395–414.

Lane, H. R., Sanderson, G. A., and Verville, G. J., 1972, Uppermost Mississippian-basal Middle Pennsylvanian conodonts and fusulinids from several exposures in the south-central and southwestern United States: 24th International Geological Congress, Montreal, 1972, sec. 7, Paleontology, p. 549–555.

Lane, N. G., 1961, An occurrence of *Archimedes* in southern Nevada: Journal of Paleontology, v. 35, p. 874–875.

———1963, A silicified Morrowan brachiopod faunule from the Bird Spring Formation, southern Nevada: Journal of Paleontology, v. 37, p. 379–392.

Langenheim, R. L., Jr., 1956, Lower Mississippian stratigraphic units in southern Nevada (abstract): Geology Society of America Bulletin, v. 67, p. 1733.

———1963, Mississippian stratigraphy in southwestern Utah and adjacent parts of Nevada and Arizona, p. 30–41. *In* Heylmun, E. G., (ed.), Guidebook to the geology of southwestern Utah: Intermountain Association of Petroleum Geologists Twelfth Annual Field Conference, 232 p.

Langenheim, R. L., Jr., and Collinson, C. W., 1963, Upper Devonian Crystal Pass Limestone of southern Nevada (abstract): Geological Society of America Special Paper 73, p. 45.

Langenheim, R. L., Jr., and Langenheim, V. A. M., 1965, The Bird Spring Group, Chesterian through Wolfcampian, at Arrow Canyon, Arrow Canyon Range, Clark County, Nevada: Illinois State Academy of Science Transactions, v. 58, p. 225–240.

Langenheim, R. L., Jr., et al., 1962, Paleozoic section in Arrow Canyon Range, Clark County, Nevada: American Association of Petroleum Geologists Bulletin, v. 46, p. 592–609.

Langenheim, R. L., Jr., et al., 1977, Fusulinids in carbonate facies of Bird Spring Group, Arrow Canyon Quadrangle, Clark County, Nevada: Journal of Paleontology, v. 51, p. 1016–1022.

Longwell, C. R., 1928, Geology of the Muddy Mountains, Nevada with a section through the Virgin Range to the Grand Wash Cliffs, Arizona: U.S. Geological Survey Bulletin 798, 152 p.

——**1936,** Geology of the Boulder Reservoir floor, Arizona–Nevada: Geological Society of America Bulletin, v. 47, p. 1393–1476.

——**1949,** Structure of the northern Muddy Mountain area, Nevada: Geological Society of America Bulletin, v. 60, p. 923–967.

——**1952,** Structure of the Muddy Mountains, Nevada. *In* Thune, H. W. (ed.), Cedar City, Utah, to Las Vegas, Nevada: Utah Geological Society Guidebook to the Geology of Utah, No. 7, p. 109–114.

——**1960,** Possible explanation of diverse structural patterns in southern Nevada: American Journal of Science, v. 258–A, p. 192–203.

——**1962,** Restudy of the Arrowhead Fault, Muddy Mountains, Nevada: U.S. Geological Survey Professional Paper 450–D, p. D82–D85.

——**1963,** Reconnaissance geology between Lake Mead and Davis Dam, Arizona–Nevada: U.S. Geological Survey Professional Paper 374–E, 51 p.

——**and Dunbar, C. O., 1936,** Problems of Pennsylvanian–Permian boundary in southern Nevada: American Association of Petroleum Geologists Bulletin, v. 20, p. 1198–1207.

Longwell, C. R., et al., 1965, Geology and mineral deposits of Clark County, Nevada: Nevada Bureau of Mines Bulletin 62, 218 p.

Lumsden, D. N., Ledhetter, M. T., and Smith, G. T., 1973, Lithostratigraphic analysis at the Bird Spring–Callville Group and Pakoon Formation (Pennsylvanian–Lower Permian), Southern Clark County, Nevada: Journal of Sedimentary Petrology, v. 43, p. 655–671.

Mamet, B. L., and Skipp, B. A., 1970, Preliminary foraminiferal correlations of early Carboniferous strata in the North American Cordillera. *In* Colloque sur la stratigraphie du carbonifère: Congrès et Colloques de l'Université de Liège, v. 55, p. 327–348.

Matthews, J. S., 1976, Paleozoic stratigraphy and structural geology of the Wheeler Ridge area, northwestern Mohave County, Arizona: Northern Arizona University unpublished M.S. thesis, 144 p.

McGovney, J. E., 1973, Fusulinid paleoecology of the Missourian of the Bird Spring Group (Pennsylvanian) of the Arrow Canyon Range, Clark County, Nevada: University of Illinois unpublished B.S. thesis, 31 p.

McKee, E. D., 1945, Oak Creek Canyon: Plateau, v. 18, p. 25–32.

——**1975,** The Supai Group–subdivision and nomenclature: U.S. Geological Survey Bulletin 1395–J, p. J1–J11.

and Gutschick, R. C., 1969, History of the Redwall Limestone of northern Arizona: Geological Society of America Memoir 114, 726 p.

——**and McKee, E. H., 1972,** Pliocene uplift of the Grand Canyon region — time of drainage adjustment: Geological Society of America Bulletin, v. 83, p. 1923–1932.

McNair, A. H., 1951, Paleozoic stratigraphy of part of northwestern Arizona: American Association of Petroleum Geologists Bulletin, v. 35, p. 503–541.

Micklin, R. F., 1967, Pennsylvanian paleontology and stratigraphy at Tungsten Gap North, Arrow Canyon Range, Clark County, Nevada: University of Illinois unpublished B.S. thesis, 69 p.

Nelson, C. J., 1973, Paleoenvironmental analysis of *Chaetetes* biostromes (Pennsylvanian) of the Arrow Canyon Quadrangle, Clark County, Nevada: University of Illinois unpublished M.S. thesis, 93 p.

Nowak, F. J., and Carozzi, A. V., 1973, Microfacies of the Upper Bird Spring Group (Pennsylvanian–Permian), Arrow Canyon Range, Clark County, Nevada: Société de Physique et d'Histoire Naturelle de Geneve Archives des Sciences, v. 25, no. 3, p. 343–382.

Pfefferkorn, H. W., 1972, Distribution of *Stigmaria wedingtonensis* (Lycopsida) in the Chesterian (Upper Mississippian) of North America: The American Midland Naturalist, p. 225–231.

Pierce, R. W., 1969, Ultrastructure and biostratigraphy of the conodonts of the Monte Cristo Group, Arrow Canyon Range, Clark County, Nevada: University of Illinois unpublished Ph. D. thesis, 113 p.

——**and Langenheim, R. L., Jr., 1972,** Mississippian (Tournaisian–Viséan) conodont zones in the Great Basin, southwestern U.S.A.: Newsletters on Stratigraphy, v. 2, p. 31–44.

Poole, F. G., Sandberg, C. A., and Boucot, A. J., 1977, Silurian and Devonian paleogeography of the western United States, p. 39–65. *In* Stewart, J. H., Stevens, C. H., and Fritsche, A. E., (eds.), Paleozoic Paleogeography of the western United States: Pacific Section, Society of Economic Paleontologists and Mineralogists, Los Angelés, 502 p.

Reinbold, M. L., and Langenheim, R. L., Jr., 1977, Late Devonian (Frasnian–Famennian) conodont biostratigraphy, Las Vegas Range, Clark County, Nevada, U.S.A.: Newsletters on Stratigraphy, v. 6, p. 183–193.

Rice, W. D., and Langenheim, R. L., Jr., 1974, (issued 1977), Conodont zonation of the Battleship Wash Formation, Late Mississippian, Arrow Canyon Range, Clark County, Nevada: Wyoming Geological Association Earth Science Bulletin, v. 7, p. 3–13.

Rich, M., 1961, Stratigraphic section and fusulinids of the Bird Spring Formation near Lee Canyon, Clark County, Nevada: Journal of Paleontology, v. 35, p. 1159–1180.

——**1971,** Middle Pennsylvanian rocks of eastern Great Basin: American Association of Petroleum Geologists Bulletin, v. 55, p. 432–453.

Robinson, H. H., 1913, The San Franciscan volcanic field: U.S. Geological Survey Professional Paper 76, 213 p.

Ryan, J. F., and Langenheim, R. L., Jr., 1973, Upper Devonian sandstone in Arrow Canyon Quadrangle, Clark County, Nevada: American Association of Petroleum Geologists Bulletin, v. 57, p. 1734–1742.

Sando W. J., Mamet, B. L., and Dutro, J. T., Jr., 1969, Carboniferous megafaunal and microfaunal zonation in the northern Cordillera of the United States: U.S. Geological Survey Professional Paper 613E, p. E1–E25.

Santogrossi, P. A., 1977, Environmental analysis of a lithostrotionid biostrome, Yellowpine Limestone, Arrow Canyon Range, Clark County, Nevada: University of Illinois unpublished M.S. thesis, 75 p.

Smosna, R. A., 1970, Tungsten Gap Chert Member, BS_c Formation, Bird Spring Group, Clark County, Nevada: University of Illinois unpublished M.S. thesis, 52 p.

Thompson, M. L., 1948, Studies of American fusulinids: Kansas University Paleontological Contributions, Protozoa, Art. 1, 184 p., 38 pls.

Tschanz, C. M., and Pampeyan, E. H., 1970, Geology and mineral deposits of Lincoln County, Nevada: Nevada Bureau of Mines Bulletin 73, 187 p.

U.S. Bureau of Reclamation, 1971, The story of Hoover Dam: 75 p.

Waines, R. H., 1961, Devonian sequence in the north portion of the Arrow Canyon Range, Clark County, Nevada (abstract): Geological Society of America Special Paper 68, p. 62.

——**1965,** Devonian stromatoporoids of Nevada: University of California at Berkeley unpublished Ph.D. thesis.

Washburn, T. A., 1968, Early Pennsylvanian crinoids from the south central Wasatch Mountains of central Utah: Brigham Young University Geology Studies, v. 15, p. 115–132.

Webster, G. D., 1968, Pennsylvanian conodonts from southern Nevada (abstract): Program, 64th Annual Meeting Geological Society of America, Cordilleran Section, p. 127.

——**1969,** Chester through Derry conodonts and stratigraphy of northern Clark and southern Lincoln Counties, Nevada: University of California Publications in Geological Sciences, v. 79, 121 p.

——**and Lane, N. G., 1967,** Mississippian–Pennsylvanian boundary in southern Nevada. *In* Teichert, C., and Yochelson, E. L. (eds.), Essays in Paleontology and Stratigraphy, R. C. Moore Commemorative Volume: Department of Geology, University of Kansas Special Publication 2, p. 502–522.

Welsh, J. E., 1960, Biostratigraphy of the Pennsylvanian and Permian Systems in southern Nevada: University of Utah unpublished Ph.D. thesis, 320 p.

White, C. A., 1874, Preliminary report upon invertebrate fossils collected by expeditions of 1871, 1872, and 1873, with descriptions of new species: U.S. Geographical and Geological Survey West of the 100th Meridian (Wheeler), 27 p.

Woodbury, D. O., 1941, The Colorado Conquest: Dodd, Mead & Company, 367 p.

DEPOSITIONAL DYNAMICS OF THE MISSISSIPPIAN REDWALL LIMESTONE IN NORTHERN ARIZONA

Richard R. Rawson[1] and W. Norman Kent[2]

Abstract — Textural analysis of the Redwall Limestone to determine facies relationships and depositional environments included classification of carbonate grain types into the following textures: early dolomite, stromatolitic wackestone, pisolitic wackestone, calcispheric packstone, pelletal packstone, oolitic packstone, oolitic grainstone, foraminiferal grainstone, crinoidal grainstone, and crinoidal packstone. These textures may be grouped into four depositional-environmental facies: Facies I, crinoidal packstone — open marine; Facies II, crinoidal and foraminiferal grainstone — open "intratidal"; Facies III, oolitic grainstone, oolitic packstone, and pelletal packstone — protected "intratidal"; Facies IV, algal (calcispheres, stromatolites, pisolites) packstone and wackestones and early dolomite — supratidal.

Facies analysis suggests two major transgressive sequences. The first major transgression spread slowly from the west but regressed rapidly. The second major transgression consisted of three rapid transgressions separated by slow regressions caused by prograding sedimentation. Facies geometry of the first transgression best fits the Shaw–Irwin model of epeiric deposition whereas the second transgression best fits Coogan's asymmetrical model of epeiric deposition.

INTRODUCTION

The Redwall Limestone in northern Arizona was deposited on a cratonic shelf during the Mississippian period. The shelf was subject to at least four transgressions and regressions. The depositional dynamics of the Redwall in northern Arizona may be demonstrated by facies analysis based on carbonate rock textures.

Previous Work

The Redwall Limestone was named by Gilbert in 1875 and originally included rocks both older and younger than Mississippian in age. Darton (1910) selected a type section in a canyon of the Shinumo drainage basin of the Grand Canyon and named the canyon "Redwall Canyon." Noble (1922) redefined the Redwall to its present configuration which includes all rocks of Mississippian age. Gutschick (1943) subdivided the Redwall into four informal members. In 1963 McKee formally named these members. McKee and Gutschick (1969) combined to write one of the most inclusive and definitive works on the Redwall. Many other paleontologists and stratigraphers, too numerous to mention here, have contributed to the understanding of the Redwall and have been recognized by McKee and Gutschick (1969). More recent work by Smith (1974), Kent (1975), and Purves (1976) has focused on facies analysis of the Redwall Limestone. The work of Smith (1974) and Kent (1975) is incorporated into this paper.

Methods

Our findings are largely the result of detailed petrologic studies of the Redwall Limestone. Detailed descriptions of carbonate textures are based on Dunham's (1962) *Classification of Carbonate Rocks According to Depositional Texture*. Textures are grouped into mappable facies using both surface and subsurface control. The Shaw (1964)–Irwin (1965) model proved to be very useful in understanding the dynamics of Redwall deposition and establishing time surfaces for constructing facies maps and interpreting the depositional history.

STRATIGRAPHY

The Redwall was deposited on a cratonic shelf that covered an area extending from the hinge line near the Nevada-Arizona border into New Mexico (fig. 65). The shelf extended through much of the Rocky Mountains during Mississippian time (Rose, 1976). The Redwall Limestone in northern Arizona includes all of the sediments deposited on this helf during the Mississippian.

At the type section, in Redwall Canyon, the Redwall overlies unconformably the Cambrian Muav Limestone, but in nearby local areas it overlies channel-fill deposits of the Devonian Temple Butte Formation. In central and northwestern Arizona the Redwall was deposited unconformably on the Devonian Martin Formation. In northeastern Arizona the Redwall overlies the Devonian Ouray Limestone.

The limestone is unconformably overlain by the Pennsylvanian Molas Formation in the subsurface of northeastern Arizona and by the Pennsylvanian Watahomigi Formation (Supai Group, McKee 1975) in central and northwestern Arizona. McKee (1963) divided the formation into four members: (in ascending order) Whitmore Wash, Thunder Springs, Mooney Falls, and Horseshoe Mesa (fig. 66). Figure 67 is an index map of the area covered by this paper and Figure 68 shows the isopach of the Redwall Limestone for the same area. The remaining isopach and facies maps also cover the same area as shown on the index map.

Whitmore Wash Member

The type section is located just north of the Colorado River in Whitmore Wash on the upthrown side of the Hurricane Fault

[1] Department of Geology, Northern Arizona University, Flagstaff, Arizona 86011.
[2] Union Oil Company, Anchorage, Alaska 99501.

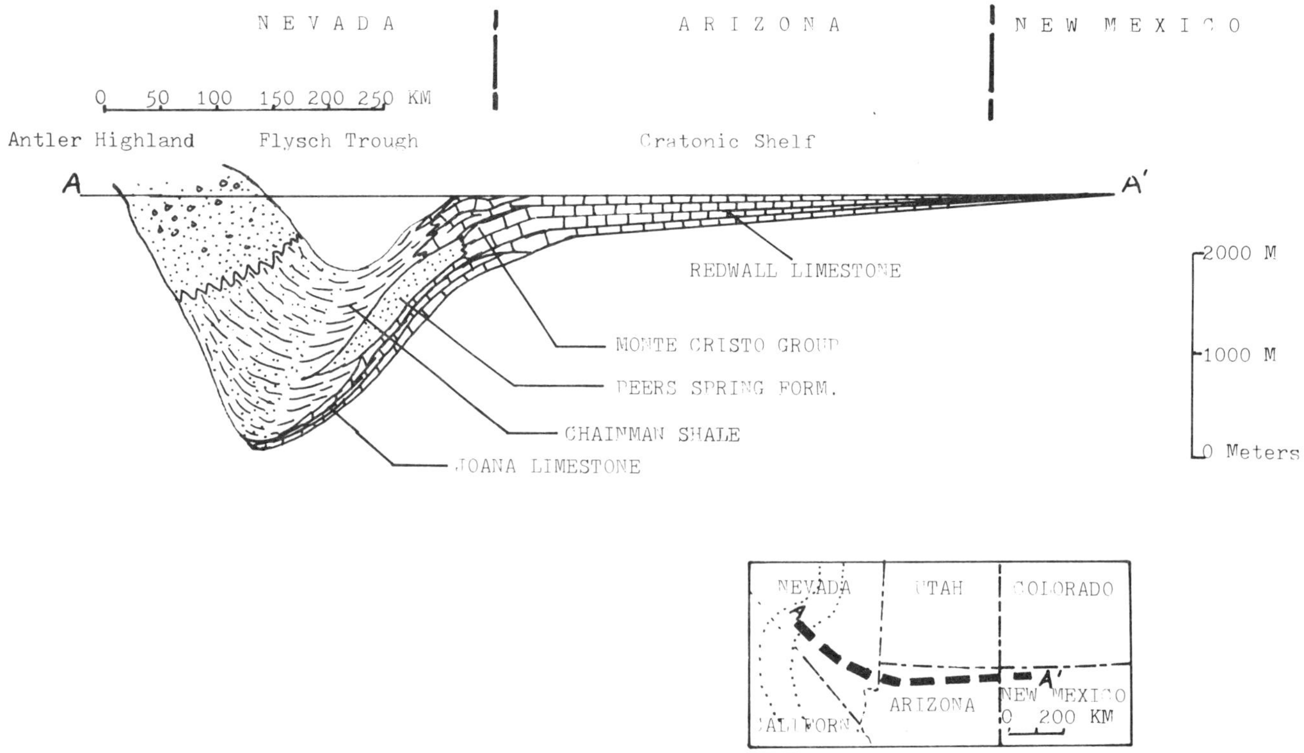

Figure 65. Cross-section from Nevada through northern Arizona into New Mexico showing relationship of Redwall Limestone to other Mississippian units to west (after Rose, 1976).

where McKee and Gutschick (1969) measured 101 feet (31 m) of thickly bedded, fine-grained dolomite. In other areas lime mudstones, pisolites, peloids, and skeletal fragments have been reported from this member (Kent 1975, Smith 1974). The Whitmore Wash ranges in thickness from an erosional zero edge in the east to a maximum thickness of over 200 feet (60 m) at Iceberg Ridge near the Airzona-Nevada border on the Colorado River (see fig. 69).

Thunder Springs Member

The type section is located at the head of Thunder River, about 2 miles north of the Colorado River in central Grand Canyon (McKee, 1963). McKee described the member as consisting of thin-bedded carbonate rock and interbedded chert. In the subsurface of northeastern Arizona this member is typically a coarse-grained late dolomite with interbedded tripolitic chert (Kent, 1975). The isopach map (fig. 70) of the Thunder Springs indicates a thickening to the north. Kent reports a thickness of 235 feet (71 m) from a well (Underwood 1–32) drilled near Jacob Lake on the Kaibab Plateau.

Mooney Falls Member

The type section is located in Havasu Canyon at Mooney Falls 4 miles south of the Colorado River on the Havasupai Indian Reservation. McKee (1963) measured 312 feet (95 m) of Mooney Falls at the type section. Kent (1975) reported approximately 450 feet (137 m) of this member in the Western #1 Valen-Fed. well located in sec. 31, T. 38 N., R. 5 W. (fig. 71). At the type section McKee and Gutschick (1969) described skeletal limestone with beds of dolomite near the base. In northeastern Arizona Kent (1975) reported an abundance of grain types including dolomite, algae, pellets, ooids, foraminifers, and crinoids. The absence of chert in the member helps distinguish it from the underlying Thunder Springs Member (Smith, 1974).

Horseshoe Mesa Member

The type section is located at Horseshoe Mesa on the south rim of the Grand Canyon below Grand View Point. McKee (1963) measured 38 feet (12 m) at the type section and described the rock as being mostly aphanitic limestone. Kent (1975) had a problem recognizing the member in the subsurface due to post-Redwall pre-Supai dissolution, weathering, and erosion removing much of the unit.

Post-Horseshoe Mesa Member

A Chesterian fauna was first reported by McKee and Gutschick in 1969 from an uppermost limestone in the Redwall near the Bright Angel Trail on the south rim of the Grand Canyon. Later work by Skipp and McKee (1978) and Billingsley (this guidebook) has verified a late Chesterian transgression that filled erosional channels in the Horseshoe Mesa Member with marine limestones containing Chesterian fossils. This transgression may correspond in time with similar transgressions that occurred in the Williston Basin forming the Big Snowy Group (Rawson, 1968) and in the Doughnut Trough of northeastern Utah and northwestern Colorado depositing the Doughnut Formation (Rose, 1976). The Chesterian channel-fill deposit in the Redwall has yet to be named.

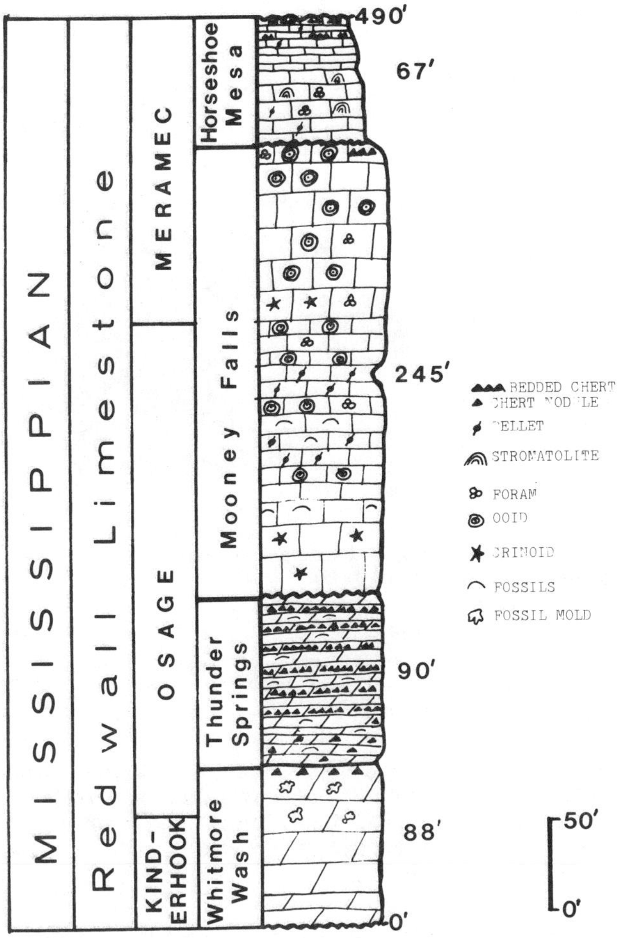

Figure 66. Redwall Limestone section on south Kaibab Trail, Grand Canyon, Arizona (after McKee, 1969, and Kent, 1975).

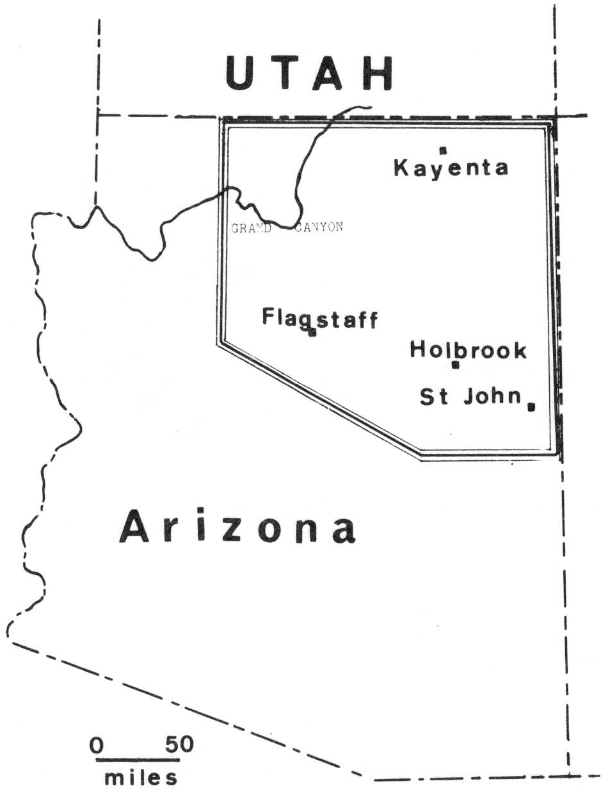

Figure 67. Index map showing area discussed in paper (after Kent, 1975).

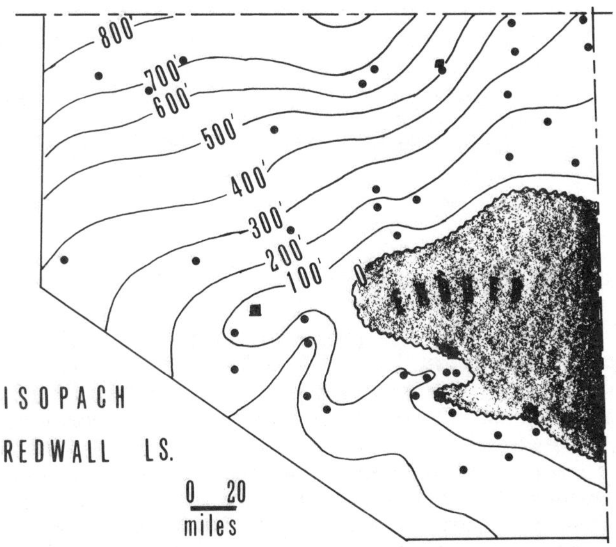

Figure 68. Isopach map of Redwall Limestone. Black circles indicate well control (after Kent, 1975).

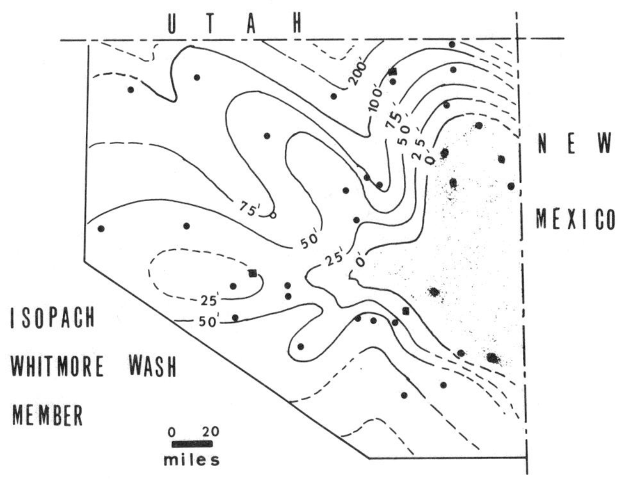

Figure 69. Isopach map of Whitmore Wash Member. Black circles indicate well control (after Kent, 1975).

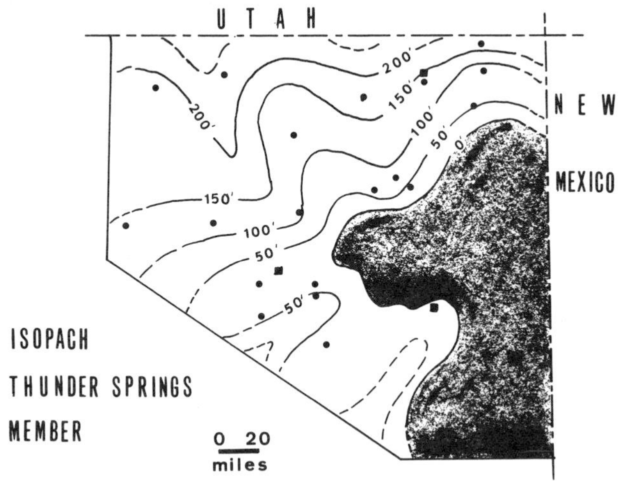

Figure 70. Isopach map of Thunder Springs Member. Black circles indicate well control (after Kent, 1975).

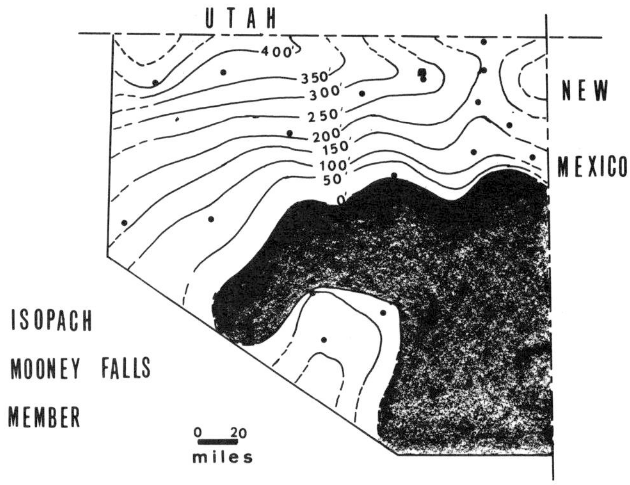

Figure 71. Isopach map of Mooney Falls Member. Black circles indicate well control (after Kent, 1975).

Age and Correlation

McKee and Gutschick (1969) concluded that the age of the Redwall ranges from late Kinderhookian to Chesterian based on fossil work done by themselves and others. The Redwall correlates with the Madison, Leadville, Escabrosa, Lake Valley, and Monte Cristo Formations or Groups of the Rocky Mountains and Great Basin.

PETROLOGY

The Redwall Limestone presents a variety of limestone and dolomite types. Smith (1974) and Kent (1975) have classified these types into two major categories: non-skeletal and skeletal. Non-skeletal types include those composed of pellets, ooids, peloids, lime mudstone, intraclasts, lithoclasts, dolomite, stromatolites, pisolites, and calcispheres. Skeletal types include crinoids, foraminifers, bryzoans, ostracodes, brachiopods, bivalves, gastropods, corals, and arthropods.

Using Dunham's (1962) classification, Kent (1975) recognized ten textures in the Redwall as having depositional significance. These include the following: early dolomite, stromatolitic wackestone, pisolitic wackestone, calcispheric packstone, pelletal packstone, oolitic packstone, oolitic grainstone, foraminiferal grainstone, crinoidal grainstone, and crinoidal packstone. Each one of these textures will be briefly described below.

Early Dolomite

Its grain size is less than 0.1 mm and is light brown to light gray in color. It may be laminated or contain peloids. This textural type is found at the base of two transgressive depositional units in the Redwall. The early dolomite is interpreted as having formed in a supratidal or sabkha environment.

Stromatolitic Wackestone

Stromatolitic structures are not abundant in the Redwall but are found at the base of the Whitmore Wash and within the Mooney Falls Members. Although widespread in several environments, stromatolitic textures are rarely preserved due to burrowing activities except in the high intertidal to supratidal environment. The stromatolitic wackestone overlies the early dolomite and is considered to have formed in or near a supratidal environment.

Pisolitic Packstone

Pisolitic packstone consists of pisolites and pellets in a matrix of micrite or neomorphic spar. Pisolites are rare in the Redwall. Where found they consist of a core of intraclasts or peloids surrounded by irregular crenulated laminations. The pisolites range in size from 2 mm to 5 mm. They appear to be algal in origin and are not similar to Dunham's (1969) vadose pisolites. The pisolitic packstone is interpreted as having formed in pools within the protected upper neritic zone in which wind-generated currents were able to roll the grains about.

Calcispheric Packstone

Calcispheres consist of a thin wall surrounding a spherical chamber filled with sparry calcite. The outside diameter is about 0.5 mm. The origin of these grains is unknown. Stanton (1963) suggests they may be plant spores or reproductive bodies while Rupp (1967) indicates that the smooth forms closely resemble the reproductive cysts of modern dasycladacian *Acetabularia*. These grains are also found in the crinoidal and pelletal packstones. This texture in the Redwall exhibits a mixture of pellets and algal structures and is interpreted as having been formed, in a lateral and vertical sense, between the pisolitic and pelletal packstones. The environment of deposition was in a low-energy neritic zone.

Pelletal Packstone

This texture consists of grain-supported pellets and fossil hash (brachiopods, ostracodes, etc.) with intergranular micrite. The pellets are roughly elliptical with no internal structure. The average diameter ranges from 0.1 mm to 0.25 mm. The texture was formed in a protected neritic, moderate-energy environment with abundant animal life. Micritic envelopes were formed on skeletal fragments associated with this texture.

Oolitic Packstone

The texture is composed of ooids, pellets, and fossil hash, including crinoids. The ooids are usually cream or yellowish white in color. Most of the nuclei are crinoid ossicles. Within the Mooney Falls Member a significant number of ooids contain foraminifers as centers. Ooid diameter in the Redwall range from 0.2 mm to 1.0 mm. Kent (1975) reported diameters up to 1.75 mm in his study of the Redwall. The oolitic packstones formed where wave energy or currents were capable of conveying ooids shelfward into a lower energy environment. The low energy of this environment is indicated by the presence of intergranular micrite.

Oolitic Grainstone

The oolitic grainstone texture is composed almost exclusively of ooids cemented by sparry calcite. Rarely, uncoated crinoids or other fossil fragments are found in association. This texture is interpreted as having formed in an environment of strong marine currents and wave action as indicated by the absence of micrite.

Foraminiferal Grainstone

Laterally and vertically the foraminiferal grainstones are found between the oolitic grainstones and the crinoidal grainstones which would suggest that foraminifers may have been able to inhabit areas of higher energy than crinoids. The types and abundance of foraminifers in the Redwall are well covered by Skipp (1969).

Crinoidal Grainstone

Crinoid ossicles represent the most abundant skeletal components of the Redwall Limestone. They range in size from 1 mm to 4 mm. The crinoidal grainstone texture may also contain an abundance of bryozoans, foraminifers, and various other types of fossil hash. The lack of micrite suggests that currents were present but not strong enough to form ooids by agitation.

Crinoidal Packstone

In this texture the components are crinoid ossicles, fossil hash, pellets, and micrite. The presence of micrite and associated components would suggest that this texture was formed seaward of the crinoidal grainstone.

FACIES ANALYSIS

From the analysis of textural types in a vertical sequence, the textures discussed above can be easily grouped into four facies that have depositional environmental significance (Kent, 1975). Walther's Law of Facies indicates that only those facies can be superimposed that can be observed laterally adjacent at the present time (Middleton, 1973). Studies of recent carbonate sediments indicate that the textures which occur vertically in the Redwall do exist adjacent to each other in modern carbonate environments. One exception would be that of the crinoids which are no longer found in shallow-water nearshore environments.

The four facies and their respective depositional textures and environments of deposition (see fig. 72) in the Redwall are: Facies I. Crinoidal packstone — open marine; Facies II. Crinoidal and foraminiferal grainstone — open "intratidal"; Facies III. Oolitic grainstone, oolitic packstone, and pelletal packstone — protected "intratidal"; Facies IV. Algal (calcispheres, stromatolites, pisolites) packstones and wackestones and early dolomite — supratidal. (The term "intratidal" as used here is defined as that area subject to wave and tidal effects but not exposed as in the intertidal zone.)

Facies I developed marineward below the wave base. Facies II and III were formed in a zone between wave base and the landward limit of the tidal effect. Facies IV formed in a zone landward from the effects of marine-generated waves and tides.

DEPOSITIONAL MODELS

Two models appear to best fit the data from the Redwall: the Shaw (1964)–Irwin (1965) epieric sea model (fig. 73) and Coogan's (1972) asymmetric carbonate cycle (fig. 74). The Redwall was deposited by two major transgressive-regressive cycles. The lower cycle deposited the Whitmore Wash and Thunder Springs Members. The Mooney Falls and Horseshoe Mesa Members constitute the upper depositional cycle. Both

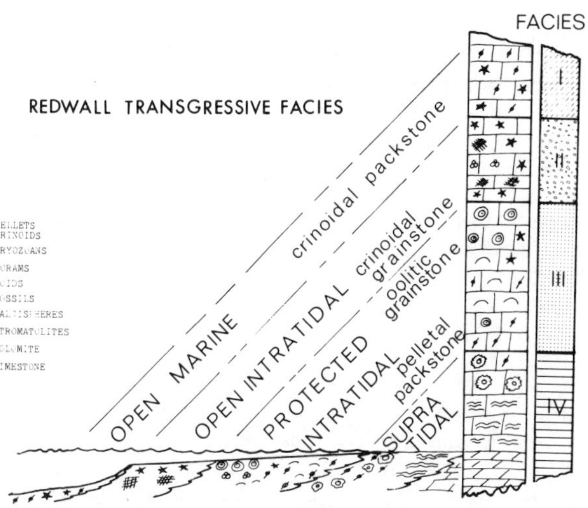

Figure 72. Relationship of depositional textures to environments of deposition and resulting facies.

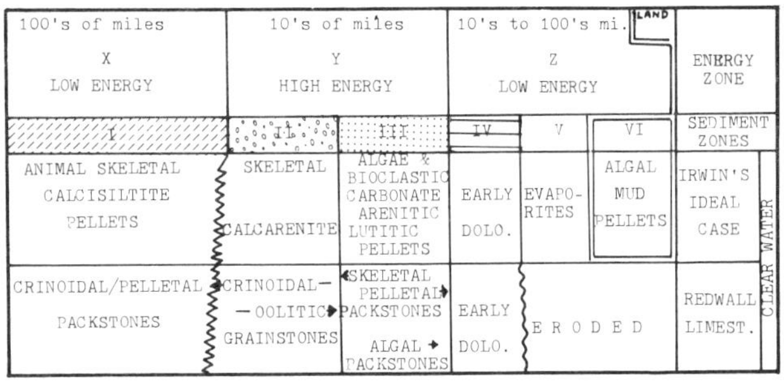

Figure 73. Comparison of Irwin's model to Redwall model (after Irwin, 1965, and Kent, 1975).

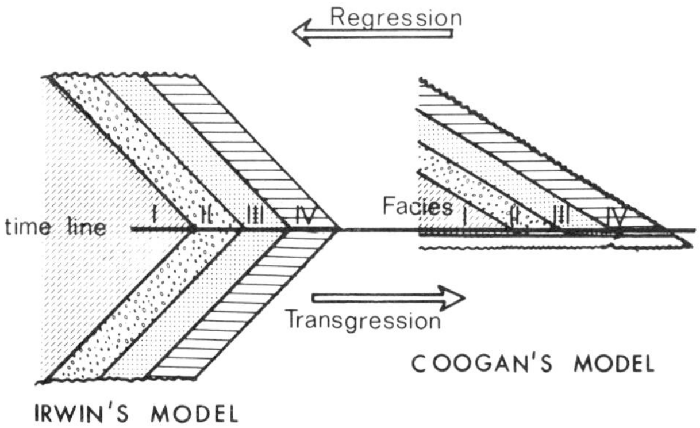

Figure 74. Comparison of Irwin (1965) and Coogan (1972) models for carbonate rocks using facies developed for Redwall Limestone.

cycles are marine and have similar facies but the manner in which the facies were displaced indicates a difference in depositional dynamics.

Lower Depositional Cycle

The transgression that occurred during the lower depositional cycle crossed a peneplane. The slope angle of the peneplane and the irregularities on that surface had a distinct effect upon the facies patterns. The steeper the slope the slower the transgression and better developed the facies. Erosional remnants, monadnocks, etc. would cause facies pattern boundaries to parallel paleotopographic contours. If the Devonian–Mississippian Ouray Limestone in northeastern Arizona is included as part of that cycle, as Kent (1975) suggested, then there was a minor regression prior to the major Redwall transgression. Once the sea had reached its maximum transgression a regression and/or progradation occurred as illustrated in Figure 75. The kickback at the base of the Redwall (R–1) and the major kickback after maximum transgression (T–1) provide two time surfaces (Irwin, 1965) upon which facies maps may be constructed (see figs. 76, 77).

The major transgression that occurred during the lower cycle was the most extensive of all the Redwall transgressions (compare fig. 77 with figs. 79, 80, 81). Throughout much of northern Arizona the Thunder Springs Member is dolomitized and the overlying Mooney Falls is not. Much of the dolomite in

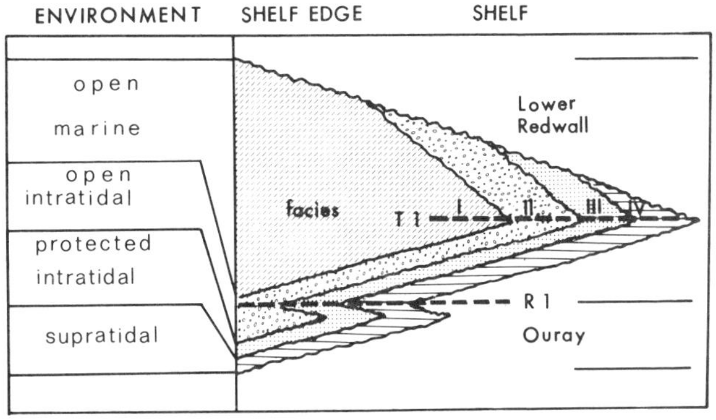

Figure 75. Cross section of idealized lower depositional cycle of Redwall Limestone showing relationship of Ouray Limestone to Redwall. Facies are shown along with two time surfaces R–1 and T–1 (after Kent, 1975).

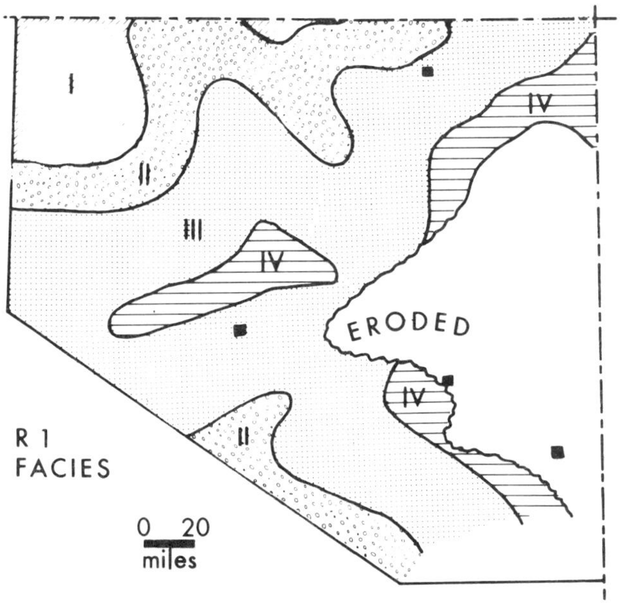

Figure 76. Facies map constructed on R–1 time surface (after Kent, 1975).

the Thunder Springs appears to be late dolomite as evidenced by the typical marine facies being dolomitized. It appears that the dolomitization occurred after the regression of the sea that deposited the Thunder Springs. Along and beneath the surface of the unconformity ideal conditions would exist to form dolomite by the mixing of marine and fresh water as proposed by Badiozamani (1973) in his Dorag dolomitization model.

Upper Depositional Cycle

The depositional textures of the Mooney Falls suggest that the transgression was fairly rapid and regression and/or depositional progradation was comparatively slow. Facies were not established until maximum transgression was attained and then as progradation of sediments occurred, accompanied by subsidence, facies were established and retreated in a marineward direction. There were at least three transgressions (fig. 78) in the upper cycle, each one overstepping the preceding one (figs. 79, 80, 81). The extent of the maximum transgression in the upper cycle (T–4) was much smaller than the one in the lower cycle (T–1) (compare fig. 77 with fig. 81).

The upper cycle of the Redwall best fits the carbonate model proposed by Coogan (fig. 75) of rapid transgression followed by slow regression and/or progradation.

SUMMARY

The Redwall Limestone in northern Arizona was deposited by an epeiric sea that covered a broad cratonic shelf. A variety of carbonate textures were formed and recorded closely their depositional environments which ranged from below-wave-base crinoidal packstone to supratidal dolomite with many intervening grain types. Facies maps reveal that the Redwall for the most part was deposited by two major transgressive-regressive cycles. The early cycle was the most extensive and appears to have been formed by a slow transgression and regression. The late cycle was marked by three fairly rapid transgressions interspersed with three slow regressions.

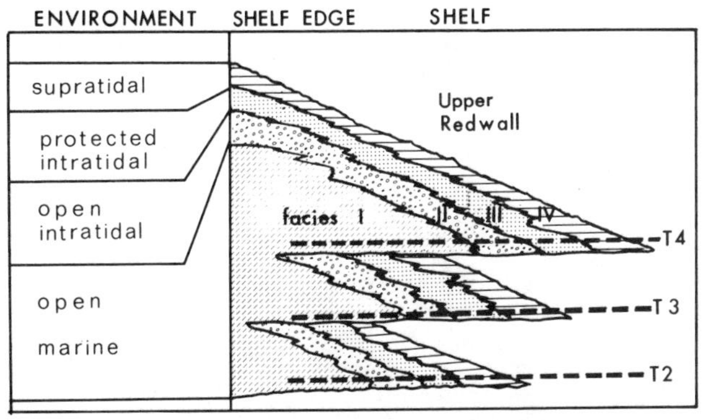

Figure 78. Cross section of idealized upper depositional cycle of Redwall Limestone showing three transgressive pulses and facies shift. Three time surfaces were constructed (T–2, T–3, T–4) just above transgressive pulse after it reached its maximum position (after Kent, 1975).

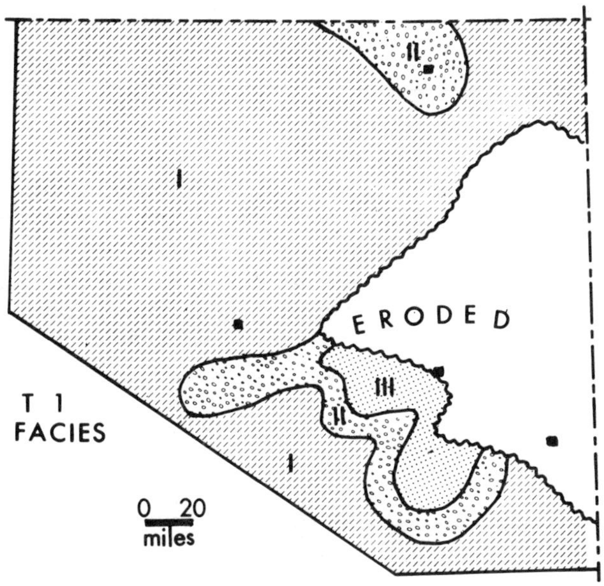

Figure 77. Facies map constructed on T–1 time surface (after Kent, 1975).

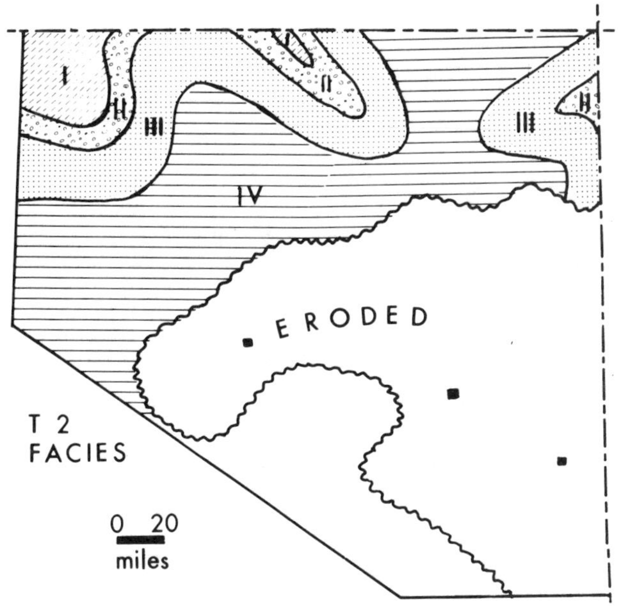

Figure 79. Facies map constructed on T–2 time surface (after Kent, 1975).

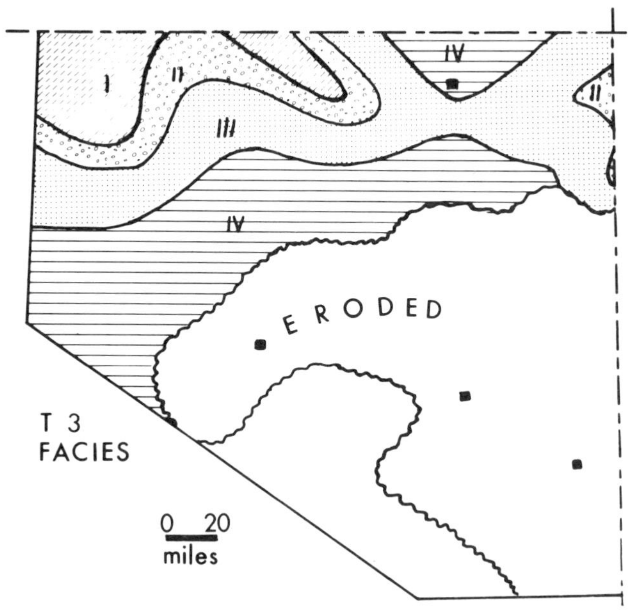

Figure 80. Facies map constructed on T–3 time surface (after Kent, 1975).

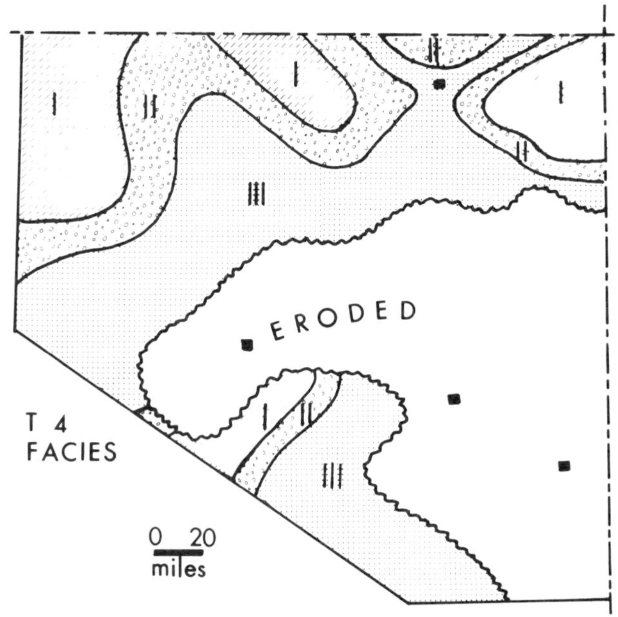

Figure 81. Facies map constructed on T–4 time surface (after Kent, 1975).

REFERENCES CITED

Badiozamani, K., 1973, The Dorag dolomitization model — application to the Middle Ordovician of Wisconsin: Journal of Sedimentary Petrology, v. 43., p. 965–984.

Coogan, A., 1972, Recent and ancient carbonate cycle sequences. *In* Elam, J. G., and Chuber, S., (eds.), Cyclic sedimentation in the Permian Basin — a symposium: West Texas Geological Society, 2nd Edition.

Darton, N. H., 1910, A reconnaissance of parts of northwestern New Mexico and northern Arizona: U.S. Geological Survey Bulletin 435, 88 p.

Dunham, R. J., 1962, Classification of carbonate rocks according to their depositional texture. *In* Ham, W. E., (ed.), Classification of carbonate rocks: American Association of Petroleum Geologists Memoir 1, p. 108–121.

———1969, Vadose pisolite in the Capitan Reef (Permian), New Mexico and Texas. *In* Friedman, G. (ed.), Depositional environments in carbonate rocks — a symposium: Society of Economic Paleontologists and Mineralogists Special Publication 14, p. 182–191.

Gilbert, G. K., 1875, Report on the geology of portions of Nevada, Utah, California, and Arizona: U.S. Geographical and Geological Survey west of 100th Meridian (Wheeler), v. 3, pt. 1, p. 17–187.

Gutschick, R. C., 1943, The Redwall Limestone (Mississippian) of Yavapai County, Arizona: Plateau, v. 16, no. 1, p. 1–11.

Irwin, M. D., 1965, General theory of epeiric clear water sedimentation: American Association of Petroleum Geologists Bulletin, v. 49, p. 445–459.

Kent, W. N., 1975, Facies analysis of the Mississippian Redwall Limestone in the Black Mesa region: Northern Arizona University unpublished M.S. thesis, 186 p.

McKee, E. D., 1963, Nomenclature for lithologic subdivisions of the Redwall Limestone, Arizona: U.S. Geological Survey Professional Paper 475–C, p. C21–C23.

———1975, The Supai Group — subdivision and nomenclature: U.S. Geological Survey Bulletin 1395–J, p. J1–J11.

———and Gutschick, R. C., 1969, History of the Redwall Limestone of northern Arizona: Geological Society of America Memoir 114, 612 p.

Middleton, G. U., 1973, Johannes Walther's law of the correlation of facies: Geological Society of America Bulletin, v. 84, p. 979–988.

Noble, L. F., 1922, A section of the Paleozoic formations of Grand Canyon at Bass Trail, U.S. Geological Survey Professional Paper 131–B, 42 p.

Purves, W. J., 1976, Possible Arizona faulting as suggested by Mississippian facies analysis and plate tectonics — a stratotectonic approach: Arizona Geological Society Digest, v. 10, March 1976, p. 259–286.

Rawson, R. R., 1968, The "Kibbey Limestone" of the Williston Basin and central Montana: Wyoming Geological Association of Earth Science Bulletin, September 1968, p. 35–37.

Rose, P. R., 1976, Mississippian carbonate shelf margins, western United States: U.S. Geological Survey Journal of Research, v. 4, no. 4, p. 449–466.

Rupp, A., 1967, Origin, structure, and environmental significance of recent and fossil calcispheres: Geological Society of America Special Paper 101, p. 186 (Abstract).

Shaw, A. B., 1964, Time in stratigraphy: New York, McGraw-Hill Book Company, 365 p.

Skipp, B., 1969, Foraminifera. *In* McKee, E. D., and Gutschick, R. C., History of the Redwall Limestone of Northern Arizona: Geological Society of America Memoir 114, p. 173–255.

———and McKee, E. D., 1978, Transgressions and regressions of Redwall Sea, Northern Arizona, related to calcareous foraminiferal faunas: Geological Society of America Abstracts with Programs, v. 10, number 3, p. 147.

Smith, J. W., 1974, The petrology of the Mississippian Redwall Limestone in northern Yavapai County, Arizona: Northern Arizona University unpublished M.S. thesis, 88 p.

Stanton, R. O., 1963, Upper Devonian calcispheres from Redwater and South Sturgeon Lake reefs, Alberta, Canada: Bulletin Canadian Petroleum Geologists, v. 11, p. 410–418.

STRATIGRAPHY OF THE SUPAI GROUP (PENNSYLVANIAN-PERMIAN), MOGOLLON RIM, ARIZONA

Ronald C. Blakey[1]

Abstract — The Supai Group of the Grand Canyon region can be correlated southeastward through nearly continuous exposures into the Mogollon Rim. The four formations of the group, Watahomigi, Manakacha, and Wescogame (all Pennsylvanian) and Esplanade Sandstone (Permian) can be recognized from Chino Point to Sedona. Although some of the contacts are difficult to pick in the field, the relationships to the type formations are close enough to justify using the terms.

The Watahomigi Formation (Morrowan–Atokan) is composed of mudstone, micritic limestone, siltstone, and sandstone and was deposited in and adjacent to marine environments. The Manakacha Formation (Atokan) consists of red sandstone, siltstone, and minor carbonate and was deposited in tidally influenced marine environments. The Wescogame Formation (Virgilian) displays ledge- and slope-forming sandstone and siltstone and probably represents paralic deposition. The Esplanade Sandstone (Wolfcampian) is chiefly ledge- and cliff-forming cross-stratified sandstone and was deposited in a complex shoreline environment.

Discrete age differences between the Supai Group to the west and Naco Group to the east indicate that the two sequences do not intertongue as previously thought. Only Virgil-aged strata cross a low arch in the Sedona area to link the two groups.

INTRODUCTION

In 1975 McKee published a brief report outlining new stratigraphic nomenclature for Pennsylvanian and Lower Permian rocks of the Grand Canyon region. He raised the Supai Formation to group status and proposed four new formal units in ascending order: the Watahomigi, Manakacha, and Wescogame Formations and the Esplanade Sandstone. The Supai Group overlies the Redwall Limestone and is succeeded by the Hermit Shale and Coconino Sandstone. Since the mid-1940's (Huddle and Dobrovolny, 1945; McKee, 1945; Hughes, 1952; Jackson, 1952; Peirce et al., 1977) most previous workers in the Mogollon Rim have assigned the entire stratigraphic interval between the Redwall and Coconino to the Supai or Supai and Naco Formations. The Hermit Shale has been assumed to pinch out in Chino Valley. This paper will present new correlations and attempt to demonstrate the relationship between McKee's four new formations of the Supai Group in the Grand Canyon and the Supai Formation of the Mogollon Rim.

The present area of study includes the western and central Mogollon Rim from Sedona to Seligman (fig. 82). In order to establish the regional correlations, sections in the western Grand Canyon region and eastern Mogollon Rim region are presented in this report.

Numerous previous papers have discussed the Supai Formation and related rocks in the Mogollon Rim region. Most papers that have discussed regional correlation of the Supai have presented rather simplified or informal correlations between the Grand Canyon and Mogollon Rim. McKee (1945) in an article on the general geology and natural history of the Oak Creek Canyon area used the term Supai Formation to include all the sedimentary units between the Redwall Limestone and Coconino Sandstone. Although he did divide the Supai into three informal subdivisions "C," "B," and "A," there was no attempt at regional correlation. Also in 1945, Huddle and Dobrovolny presented correlations of Paleozoic rocks in central and eastern Arizona in which the three divisions of McKee (1945) were informally called the lower, middle, and upper members of the Supai Formation. Correlating eastward from Oak Creek, they showed the Supai Formation overlying the Naco Formation at Fossil Creek.

Hughes (1952) suggested that the Hermit Shale pinched out in western Chino Valley corroborating the earlier general usage of the term Supai for the entire Pennsylvanian and Permian interval below the Coconino. He had difficulty correlating his Supai Formation in the Chino Valley and Big Black Mesa areas, even between fairly closely spaced measured sections. No external correlation was attempted. Jackson (1952) correlated the entire Supai and Naco interval of the Mogollon Rim from Fort Apache to southern Nevada. Although general in nature, he clearly showed the pinchout of the Hermit Shale in western Chino Valley, the correlation of the Naco Formation westward into Nevada, and the direct correlation of the "type" Grand Canyon Supai Formation with the Supai Formation of the Mogollon Rim. None of these conclusions are accepted in this paper.

Baars (1962) restricted the term Supai in the Mogollon Rim to the lower part of the sequence. Although he was uncertain as to the correlation of Hermit Shale and it is not clear exactly which interval of the Mogollon Rim Supai he correlated to the Grand Canyon, this was a significant departure from previous interpretations.

In 1973 Ross correlated fossiliferous Pennsylvanian and Permian sequences of southeastern Arizona into the Mogollon Rim. The terms Horquilla and Earp Formations were brought into the Mogollon Rim from the southeast and placed in the Naco Group. The overlying redbeds were referred to as Supai Formation but no significant study or correlation of the Supai interval was made.

Preliminary conclusions concerning the rocks discussed in this paper were first presented by Lane (1977). The "C" member of McKee was correlated with the type Supai; the "B" member was correlated with the Hermit Shale; and the "A" member was postulated to be post-Hermit pre-Coconino and thus not present in the Grand Canyon.

Peirce et al. (1977) reviewed previous work and concluded that the divisions "A," "B," and "C" were the most workable. Their correlations follow closely the correlations of Huddle and

[1] Department of Geology, Northern Arizona University, Flagstaff, Arizona 86011.

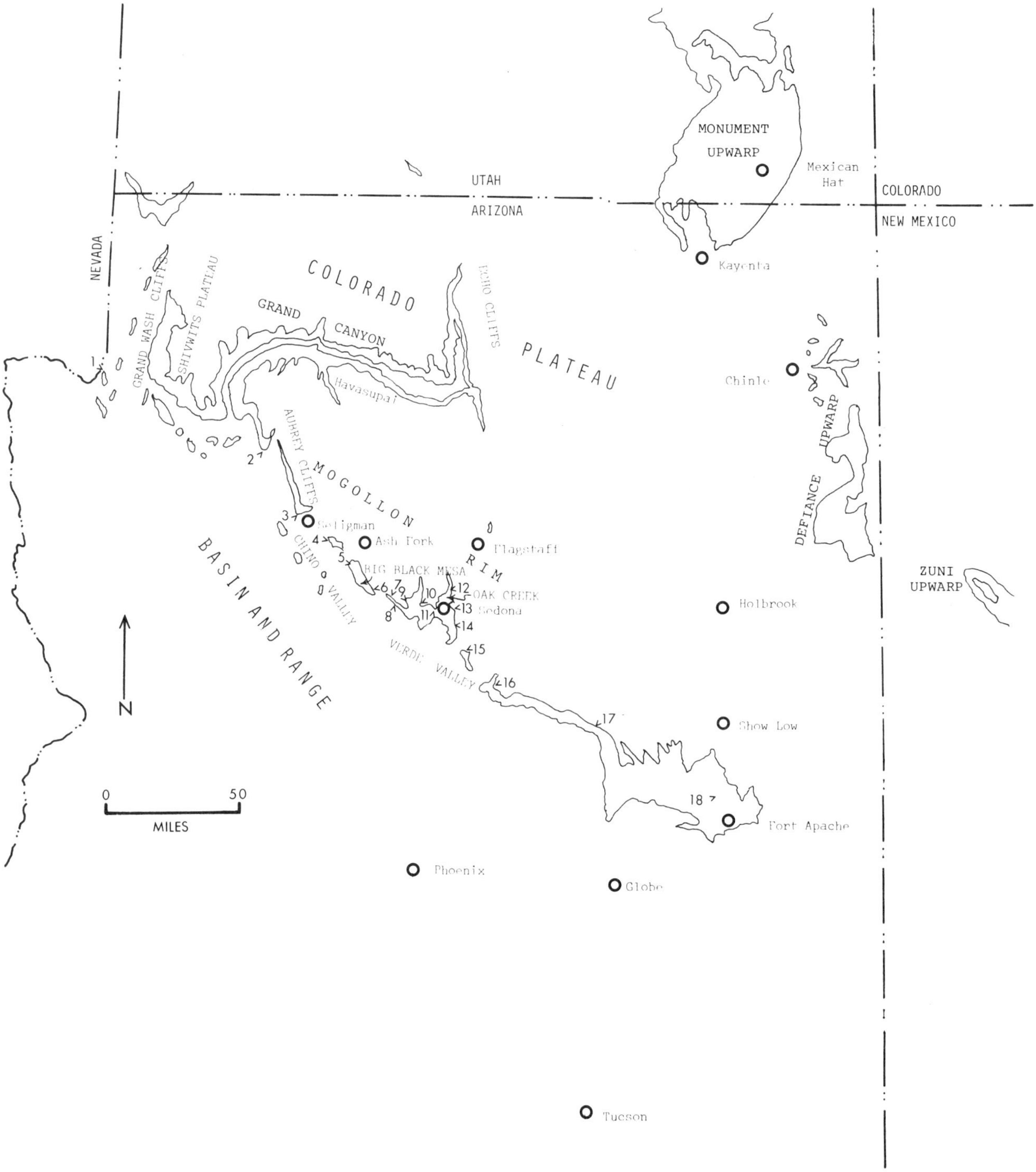

Figure 82. Map of southern Colorado Plateau and adjacent areas showing outcrop of Pennsylvanian and Lower Permian rocks. Numbered measured sections include: 1) Iceberg Ridge; 2) Blue Mountain (Diamond Creek); 3) Chino Point; 4) Picacho Butte; 5) Cathedral Caves; 6) Hell Canyon; 7) U.S. Mine; 8) West Perkinsville; 9) Perkinsville; 10) Sycamore Canyon; 11) Dry Creek; 12) Midgley Bridge; 13) Schnebly Hill; 14) Bell Rock; 15) Beaver Creek; 16) Fossil Creek; 17) Arizona Highway 260; 18) Corduroy Creek (Fort Apache).

Dobrovolny (1945) and Jackson (1952); the work of Baars (1962) and Ross (1973) was not followed.

At present it is obvious that disagreement exists as to whether and how the Supai of the Mogollon Rim correlates to the Supai of the Grand Canyon. Few of the previous studies used detailed faunal evidence which has only recently become available.

REGIONAL CORRELATION

Tectonics

Regional correlation of the Supai Group becomes clearer if the regional tectonics are understood. Blakey (1978) emphasized the tri-basin aspect of the Pennsylvanian of the southern Colorado Plateau (fig. 83). Southeastern Arizona was the site of the Pedregosa Basin; northeastern Arizona and southeastern Utah were the site of the Paradox Basin; northwestern Arizona and adjacent Nevada and Utah were the site of the Cordilleran Miogeosyncline. Each basin had an independent history; intervening shelves, arches, and upwarps separated the basins. Most of the Supai Group was deposited on a shelf and embayment. The former was termed the Mogollon Inner Shelf by Ross (1973) and the latter the Grand Canyon Embayment by McKee (1978, personal commun.). The Supai Group was deposited on the edge of the Cordilleran Miogeosyncline mostly within the Grand Canyon Embayment. Most of the area along the Mogollon Rim was the site of the Mogollon Inner Shelf. The Naco Formation (Group) was deposited within the Pedregosa Basin and on the eastern portion of the Mogollon Inner Shelf. The Hermosa Group was deposited in the Paradox Basin and surrounding areas.

Correlation of the type Supai Group

The first attempt at division and correlation of the Grand Canyon Supai was based on topographic expression (McKee, 1975). The formations could easily be divided into cliffs and slopes, most of which correlated throughout the Grand Canyon. McKee found that these correlations were supported by fossils, especially fusulinids, and he established the four new formations on this basis (fig. 84).

On the Mogollon Rim, most attempts at subdivisions were based on the tripart divisions of McKee (1945) and Huddle and Dobrovolny (1945). In general, the lowest division comprises cliff-, ledge-, and slope-forming sandstone, siltstone, and carbonate. The entire sequence is limey and generally composed of pink, orange, gray, and red units. The age has generally been considered Pennsylvanian with possible Permian near the top (Peirce et al., 1977). The middle subdivision consists of slope-forming siltstone, sandstone, conglomerate, and silty carbonate. The sequence is very limey and generally pale reddish brown in color. The age has generally been assumed to be Pennsylvanian and Permian. The third subdivision comprises cross-stratified sandstone to the west and siltstone, sandstone, limestone, dolomite, and gypsum to the east. It forms towering cliffs and slickrock ledges to the west and steep slopes to the east. Some cross-stratified sandstone is noncalcareous and the sequence is generally reddish orange in color. Several carbonates, most prominently the "Fort Apache Limestone Member of the Supai Formation," contain fossils of early Leonard (Permian) age (Gerrard, 1966).

Previous regional correlations of the Supai Formation of northern Arizona, including the restriction of Supai Formation on the Mogollon Rim by Baars (1962), have concentrated on the eastern Mogollon Rim. These correlations have been hindered by tenuous subsurface correlations involving unfossiliferous rocks; rocks along the Mogollon Rim were correlated to better known sections in New Mexico, Utah, and northeastern Arizona. Unfortunately the type Supai and closely equivalent rocks are not present in these areas.

This work correlates the Supai from the western and central Grand Canyon southeastward into the western and central Mogollon Rim (fig. 84). Extensive subsurface correlation is not necessary because the largest gap in outcrop is only 15 miles (24 km) and certain key units used in the correlation are exposed in a nearly continuous outcrop from Diamond Creek in the western Grand Canyon to Oak Creek Canyon at Sedona.

McKee (1978, personal commun.) has correlated the type Supai Group from Havasupai to Diamond Creek. The four formations can then be correlated to Picacho Butte, through Big Black Mesa, into the Verde Valley near Perkinsville, and eastward to Dry Creek. At all localities, the Esplanade Sandstone marks the top of the group. Numerous partial sections at localities between the above-mentioned places substantiate the correlations. From Chino Valley eastward, the Esplanade Sandstone underlies what has been previously called the top of the Supai Formation, and at Midgley Bridge in Oak Creek Canyon, the top of the Esplanade is some 1,200 feet (365 m) below the Coconino Sandstone. At several scattered locations in Chino and western Verde Valleys, a slope-forming red sandstone and siltstone sequence overlies the Esplanade. This in turn lies between the Esplanade and the towering red-orange cliffs in Verde Valley (fig. 85). The former has been called the middle or "B" member of the Supai Formation and the latter the upper or "A" member (McKee, 1945; Huddle and Dobrovolny, 1945; Lehner, 1958; Krieger, 1965; Peirce, et al., 1977). The slope-forming "middle member" is the Hermit Shale and the cliff-forming upper member is a unit not represented in the Grand Canyon but deposited on the edge of and within the Permian Holbrook Basin. This unit is informally named the Schnebly Hill formation after excellent exposures at Schnebly Hill several miles east of Sedona (fig. 86). The remainder of this paper will support and further define these correlations.

Definition and Correlation of the Supai Group, Mogollon Rim

The Esplanade Sandstone is the only formation of the Supai Group that can be easily recognized, mapped, and correlated into the Mogollon Rim. The cliff and slope topography and excellent continuous exposures of the Supai Group that aided McKee (1975) in the Grand Canyon are not as well developed (fig. 87). Several places in Sycamore Canyon display similar outcrop expression as at the Grand Canyon but in general, the units underlying the Esplanade display a somewhat different topographic expression. In general, throughout the west-central Mogollon Rim from Sycamore Canyon to Picacho Butte, the following subdivisions can be recognized, in ascending order: 1) cherty limestone, silty limestone and siltstone that form gray slopes and ledges, 2) strongly cyclic nodular silty limestone, micritic limestone, siltstone, and bioturbated sandstone that form orange and brown ledges and slopes, 3) cross-stratified, reddish-orange sandstone that forms a persistent cliff or broken ledge, 4) cross-stratified to bioturbated, limey sandstone and limey siltstone, reddish-orange to pale grayish orange, that form ledges and cliffs, and 5) pale grayish-orange, cross-stratified sandstone that forms cliffs. Unit 1 and most of unit 2 belong to the Watahomigi Formation; a small portion of unit 2, unit 3, and part of unit 4 belong to the Manakacha Formation; most of unit 4 belongs to the Wescogame Formation; and unit 5 belongs to the Esplanade Sandstone.

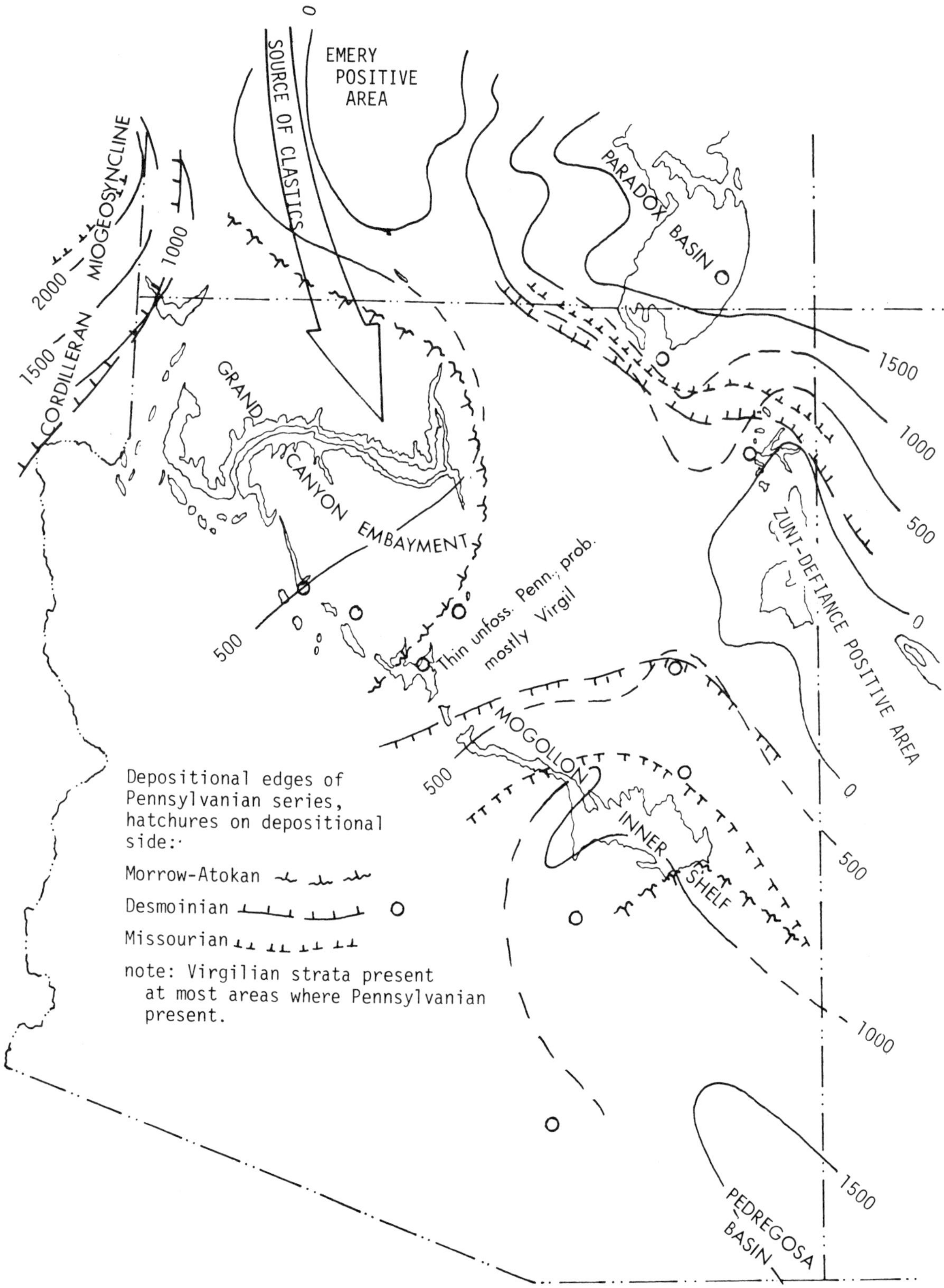

Figure 83. Isopach map of Pennsylvanian rocks on southern Colorado Plateau and adjacent areas. Note distinct tri-basin aspect. Open circles represent towns shown on Figure 82. Isopachs given in feet.

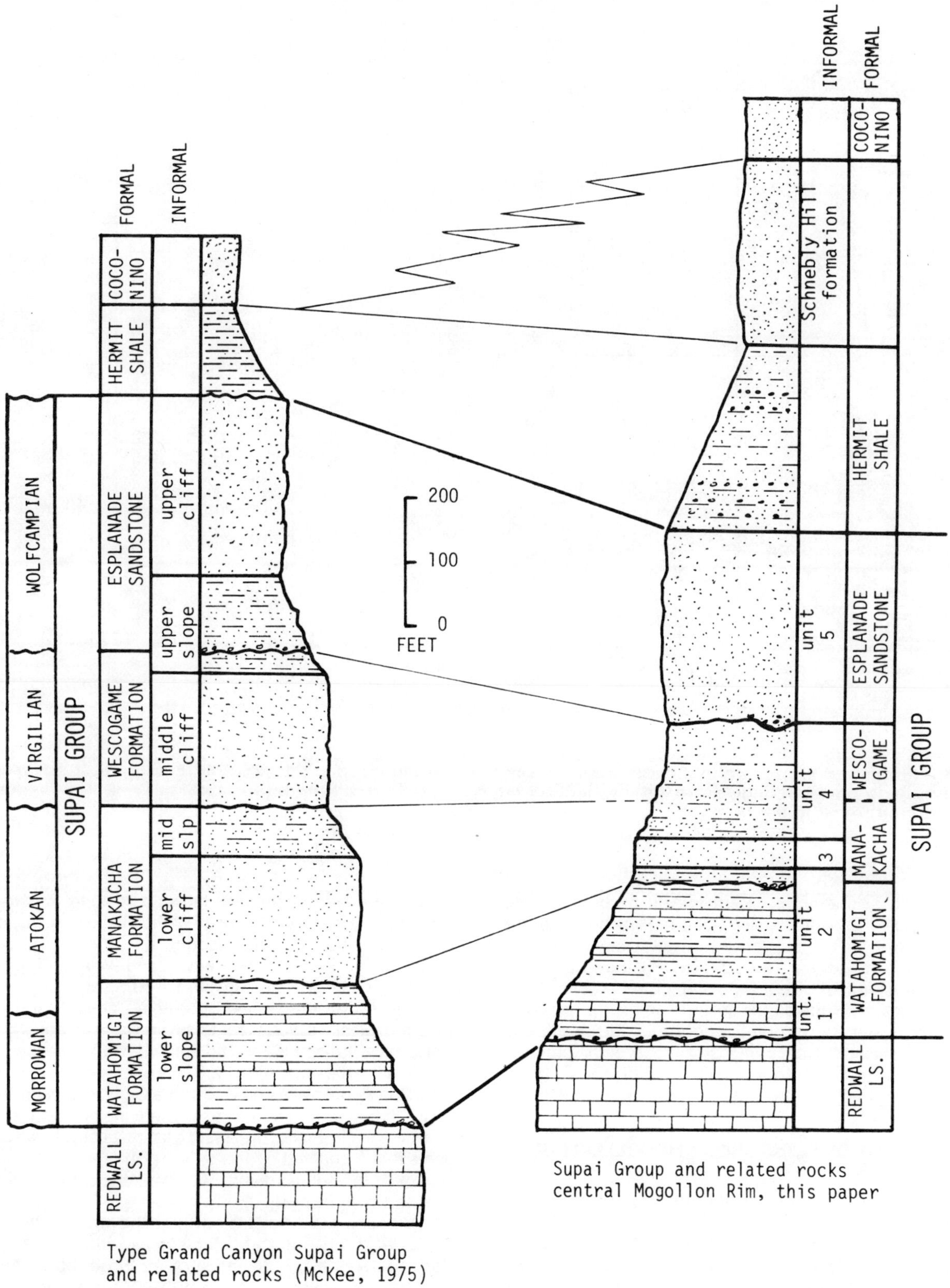

Figure 84. Columnar sections of type Supai Group, Grand Canyon, and Supai Group and related rocks, Mogollon Rim, showing correlations between the two regions.

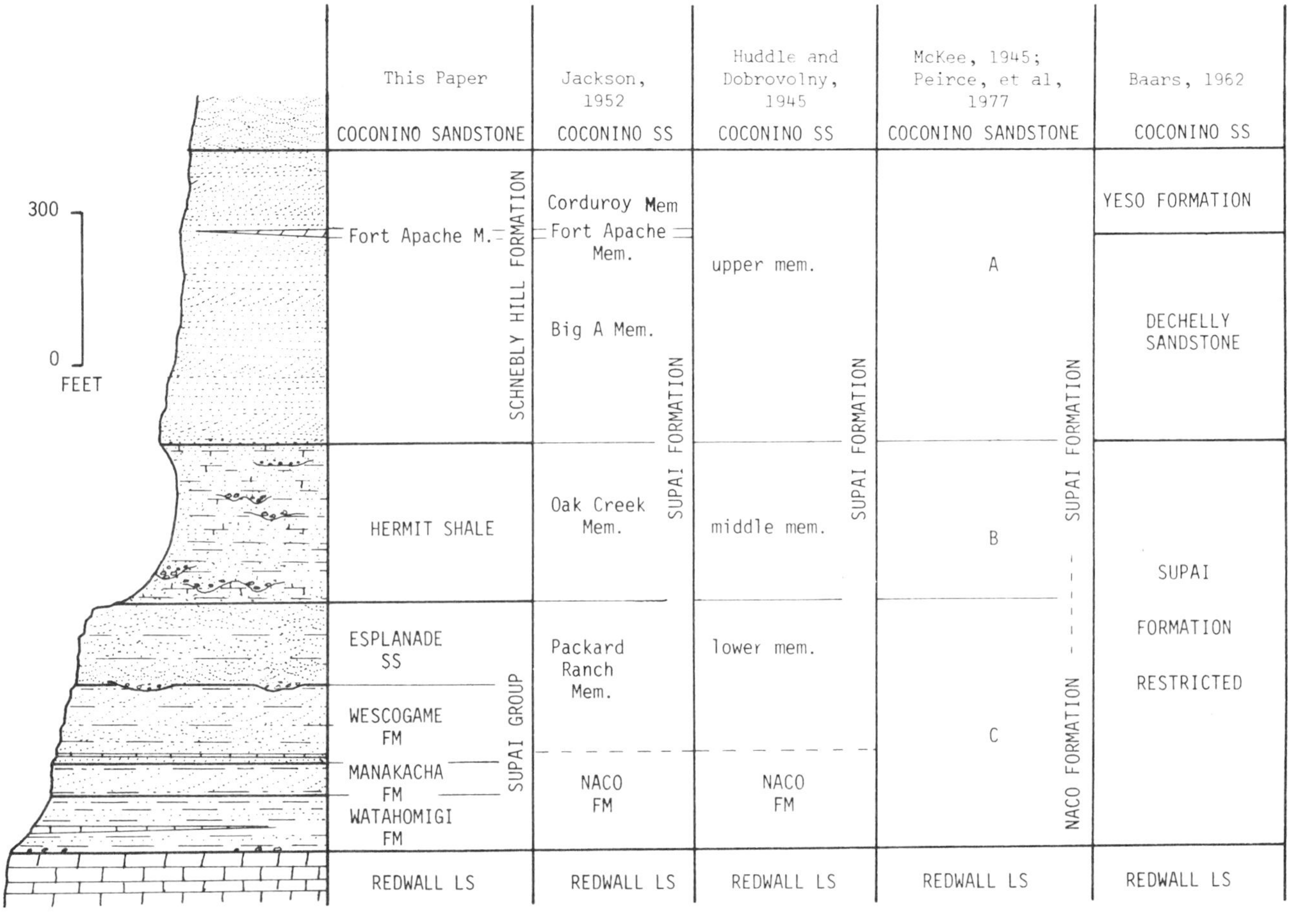

Figure 85. Generalized columnar section of Pennsylvanian and Lower Permian rocks in central Mogollon Rim showing present and previous nomenclature. The term Schnebly Hill formation is informal in this report. Baars (1962) correlated upper portion of Supai Formation with Organ Rock Shale which in turn correlates with Hermit Shale.

The overlying slope-forming Hermit Shale can be correlated from the Aubrey Cliffs and Chino Point to Picacho Butte, the Western Verde Valley, Sycamore Canyon, Oak Creek, and eastward to Fort Apache. The Schnebly Hill formation thickens to the east from a zero edge along a line between Chino Point and Picacho Butte on the south to an area a few miles east of the Grand Canyon on the north (fig. 88). Neither of these units, which together comprise the bulk of the previous "Supai Formation" in the Mogollon Rim, are discussed in detail here.

STRATIGRAPHY OF SUPAI GROUP, MOGOLLON RIM

Underlying Rocks and Basal Unconformity

At all locations on the Mogollon Rim, the Supai Group, as defined here, overlies the Redwall Limestone of Mississippian age. The massive-weathering cliffs, rounded ledges, and knobby, partially covered slopes of the Redwall contrast with the thinner cliffs and ledges and chert-covered slopes of the overlying Watahomigi Formation. The similar-appearing, fine-grained, medium-gray Mississippian and Pennsylvanian carbonates can be difficult to distinguish in hand specimen or in areas of poor exposure but in all locations, except possibly Picacho Butte, the basal Watahomigi carbonate is separated from the Redwall by purplish-brown mudstone. A more complete discussion of the lithology of the Redwall is provided by Rawson and Kent (this guidebook).

The contact between the Redwall and Watahomigi is poorly exposed. Only at Hell Canyon, the U.S. Mine, and several areas near Perkinsville and Sycamore Canyon is the contact visible. Typically several feet to several tens of feet of covered slope mark the boundary. The amount of relief on the contact is difficult to measure directly. Basal chert-pebble conglomerate in the Perkinsville and Sycamore Canyon sections ranges up to 30 feet (9 m) in thickness. The conglomerate is absent in sections a few miles west and east respectively, which suggests that a channel on the order of 30 feet (9 m) deep was filled with the conglomerate. At numerous localities a poorly exposed but obviously undulating surface marks the boundary. Although it is difficult to estimate the relief on this surface it is at least 3 to 5 feet (1-1.5 m) locally and may be greater. Karst features and surfaces such as those reported farther east in the Mogollon Rim by Brew (1965) are not as obvious in the central and western Mogollon Rim, although several areas of karst topography were reported by McKee and Gutschick (1969).

The span of time missing at the Mississippian-Pennsylvanian boundary in the Mogollon Rim probably ranges from early to middle Meramecian to middle Morrowan. McKee and Gutschick (1969) recognized the Mooney Falls Member as the up-

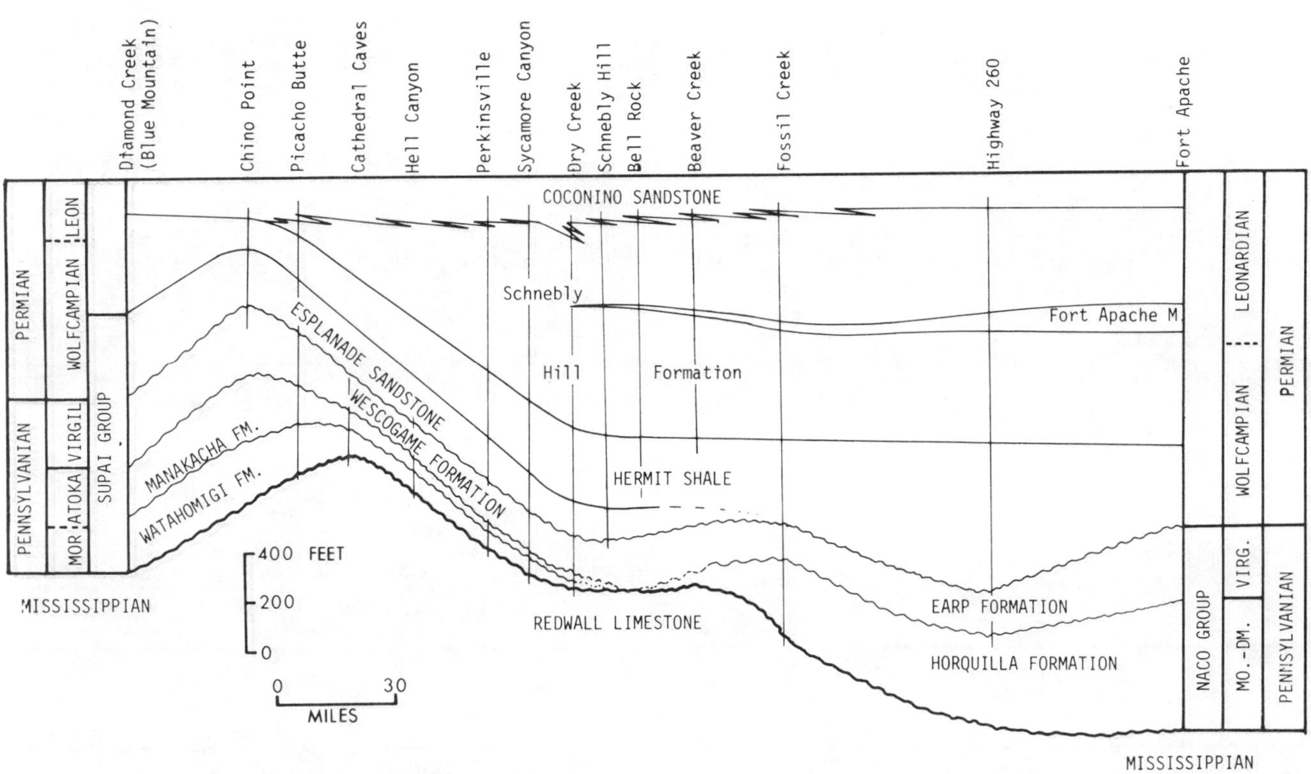

Figure 86. Terminology of Pennsylvanian and Lower Permian rocks of Mogollon Rim as used in this report. The Schnebly Hill formation is informal in this report. Vertical lines show interval measured for each locality.

permost unit in the Redwall Limestone in most of the Mogollon Rim and assigned it an Osagian and possible early Meramecian age. Gordon and McKee (1978) assigned a middle upper Morrowan age to the oldest fossiliferous units of the Watahomigi Formation in the Grand Canyon. A fossiliferous carbonate in the Watahomigi on the Mogollon Rim at Cathedral Caves is probably of the same age. Hughes (1952) cited evidence that the fauna is Early Pennsylvanian. This same limestone can be traced at least as far east as Perkinsville although it is unfossiliferous in that area. At Fossil Creek the oldest fossiliferous Pennsylvanian rocks are Desmoinesian and farther east in the Fort Apache region, Atokan (Brew, 1965; Ross, 1973). Morrowan rocks are reported in southeastern Arizona by Ross (1973). The length of time represented by the Mississipian–Pennsylvanian unconformity decreases toward the Cordilleran Miogeosyncline and Pedregosa Basin respectively.

Basal Carbonate and Conglomerate (Watahomigi Formation)

The basal sequence of the Supai Group in the Mogollon Rim is comprised of cherty carbonate, conglomerate, and fine-grained clastics. Although a gap of about 35 miles (55 km) separates outcrops in the Grand Canyon from outcrops in the Mogollon Rim, lithology, fossils, stratigraphic position, and nature of outcrop leave little doubt that the oldest rocks of the Supai Group on the Mogollon Rim correlate with the Watahomigi Formation. In areas of fair to good exposures between Picacho Butte and Perkinsville, the Watahomigi constitutes an easily mappable unit although the exact boundary with the overlying Manakacha Formation is usually difficult to pick.

Lithologically the Watahomigi comprises cherty, micritic to recrystallized, medium- to dark-gray, locally oolitic and peloidal limestone and dolomite; laminar to nodular limey, locally cherty reddish-brown to grayish-red siltstone; reddish-brown to purplish-gray homogeneous dolomitic siltstone; fine-grained, well-sorted, bioturbated, locally cross-stratified, quartzose sandstone; poorly exposed, angular, sandy, reddish, chert-pebble conglomerate; and poorly exposed, purplish siliceous mudstone and claystone. The basal portion of the formation typically forms drab-colored chert-covered slopes and ledges and the upper portion forms reddish-brown slopes and ledges. Talus from overlying units covers some of the formation particularly where the Supai outcrop belt is narrow and steep.

The Watahomigi Formation thickens westward from 63 feet (19 m) at Dry Creek to 257 feet (78 m) at Picacho Butte (fig. 89). McKee (1978, personal commun.) reported 174 and 304 feet (53 and 92 m) at Blue Mountain (Diamond Creek) and Iceberg Ridge respectively. As previously described, the Watahomigi Formation overlies the Redwall Limestone

Figure 87. Photographs of Supai Group in Mogollon Rim. A – upper Watahomigi Formation, Manakacha Formation, and lower Wescogame Formation, U.S. Mine section; B – lower two-thirds of Supai Group at U.S. Mine Section (Redwall Limestone underlies tree-covered surface along Verde River); C– upper Manakacha Formation, Wescogame Formation, and probable lowest Esplanade Sandstone at Hell Canyon section; D– most of Supai Group near Perkinsville. Symbols: PPwa– Watahomigi; PPm– Manakacha; PPwe– Wescogame; Pe– Esplanade.

disconformably. The upper boundary is marked at the base of a prominent limestone-pebble conglomerate at Perkinsville and by poorly exposed and defined erosion surfaces elsewhere. For example, at Picacho Butte an erosion surface at the base of the Manakacha cliff with some 20-25 feet (6-7 m) of relief marks the upper boundary; in Dry Creek a persistent erosion surface with several feet of local relief probably marks the boundary. Between Picacho Butte and Perkinsville, the top of the Watahomigi is placed at the base of a nodular limey sequence 10-25 feet (3-7 m) below the Manakacha cliff. For mapping purposes the boundary can be placed at the base of the Manakacha cliff. It should be pointed out that McKee (1978, personal commun.) has indicated that the Manakacha Formation displays many small unconformities at some locations in Grand Canyon and that the unconformity picked as the top of the Watahomigi on the Mogollon Rim might actually be correlated to an unconformity in the Manakacha of the Grand Canyon. Thus the position of the upper boundary on the Mogollon Rim probably marks the highest possible upper boundary for the Watahomigi.

Westward the Watahomigi Formation correlates with the lower member of the Callville Limestone in southern Nevada. McKee and his coworkers (1978, personal commun.) can apparently correlate the Watahomigi interval some distance into southern Nevada based on faunal, lithologic, and stratigraphic data. Eastward the edge of the Watahomigi Formation cannot be precisely located. Rocks of similar age are missing at Fossil Creek. Although scattered, local, thin, unfossiliferous strata may be present in the subsurface east of Grand Canyon, the term Watahomigi Formation has little meaning east of a line from the Echo Cliffs to Sedona. Morrowan rocks are absent in the eastern Mogollon Rim (Ross, 1973; Brew, 1965).

Within the west-central Mogollon Rim, the Watahomigi Formation comprises several wide-spread sequences. The base is nearly everywhere a covered slope that, where exposed, consists of a basal chert-pebble conglomerate overlain by purplish mudstone. This is succeeded by a ledge- or cliff-forming sequence of cherty limestone. The upper Watahomigi comprises reddish-brown clastics and thin micritic and locally cherty carbonate. West of Perkinsville all three sequences are present. Eastward to Dry Creek, the reddish dolomitic siltstone and sandstone probably correlates to the upper division.

The age of the Watahomigi in the Grand Canyon is middle and late Morrowan and early Atokan (Gordon and McKee, 1978). A similar age is assigned on the Mogollon Rim based on similar stratigraphic position, lithology, and fossils.

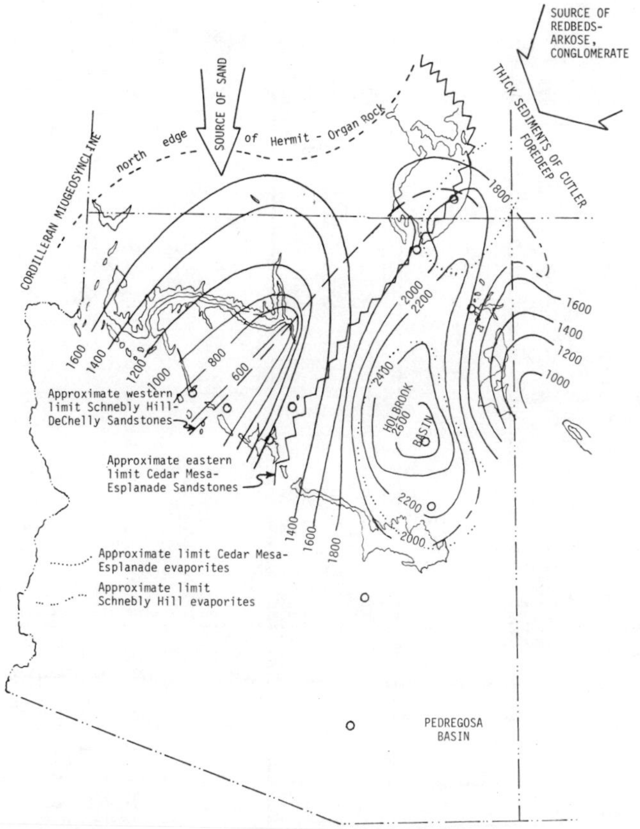

Figure 88. Isopach map of pre-Coconino Permian rocks on southern Colorado Plateau. In southeastern Utah included formations are the Halgaito, Cedar Mesa, and Organ Rock; in eastern Arizona the De Chelly Sandstone and related rocks; in central and western Mogollon Rim the Esplanade, Hermit, and Schnebly Hill; in Grand Canyon the Esplanade and Hermit. Isopachs given in feet.

Sandstone and Siltstone Sequence (Manakacha and Wescogame Formations)

The Watahomigi Formation is succeeded by a strongly cyclic sequence of reddish-orange sandstone and siltstone. This sequence correlates to the Manakacha and Wescogame Formations in the type Supai Group. The units are discussed together here because they have similar lithology and cyclic sequences and are very difficult to separate in the field. It is unlikely that the two formations could be mapped individually, at least consistently, along the Mogollon Rim; however, the dual sequence can be mapped and distinguished from the underlying and overlying units.

Both the Manakacha and Wescogame Formations consist of cross-stratified, fine-grained, orangish-gray quartzose sandstone; limey, nodular to platey, sandy, reddish-brown to grayish-red siltstone; micritic, irregularly laminated (probably stromatolitic), medium-gray dolomite and limestone; and bioturbated, massive-weathering, fine-grained sandstone. A general but persistent trend throughout the western Mogollon Rim is the gradual increase upward in sandstone at the expense of the other lithologic types.

The sequence is strongly cyclic throughout. A typical cycle includes basal cross-stratified sandstone grading upwards to, though locally succeeded abruptly by, "massive" bioturbated sandstone. This may be overlain by limestone, nodular siltstone, laminated siltstone, or all three, and is then succeeded by the overlying cycle. Ten to 15 cycles can be recognized at most localities.

Several types of cross-stratification are present. Wedge-planar and tabular-planar types dominate but trough sets are also present. Sets commonly range from 1 to 4 feet (0.3-1.0 m) thick and have dominant southerly dip components. Some units near the top of the sequence display intraset or compound cross-stratification, and near the middle of the sequence herringbone pattern occurs in several units. Some of the cross-stratified "sandstones" contain abundant carbonate grains such as oolites and fossils and would probably be classified as limestones.

Careful comparison and correlation of the Mogollon Rim sequence with the type Supai in the Grand Canyon allows a rather crude division to be drawn between the Manakacha and Wescogame Formations. Although difficult to pick in the field, the boundary appears somewhat consistent as shown in Figure 90. Based on this boundary, the Manakacha Formation thickens from 35 feet (10 m) at Dry Creek to 70–90 feet (20–28 m) in the Western Mogollon Rim (fig. 91). McKee (1978, personal commun.) recognized 267 feet (81 m) and 277 feet (84 m) at Blue Mountain and Iceberg Ridge respectively. Thickness of the Wescogame ranges from 115 to 150 feet (35–45 m) in Chino and Verde Valleys. McKee reported 102 feet (31 m) at Blue Mountain and 180 feet (55 m) at Iceberg Ridge.

The boundary between the Wescogame and Esplanade is a regional disconformity that marks the Pennsylvanian–Permian boundary. On rugged, irregular, broken cliffs the relief

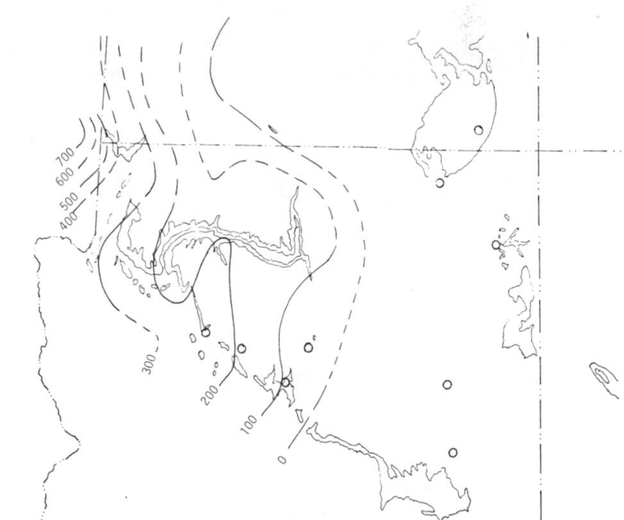

Figure 89. Isopach map of Watahomigi Formation, Southern Colorado Plateau. Isopachs given in feet.

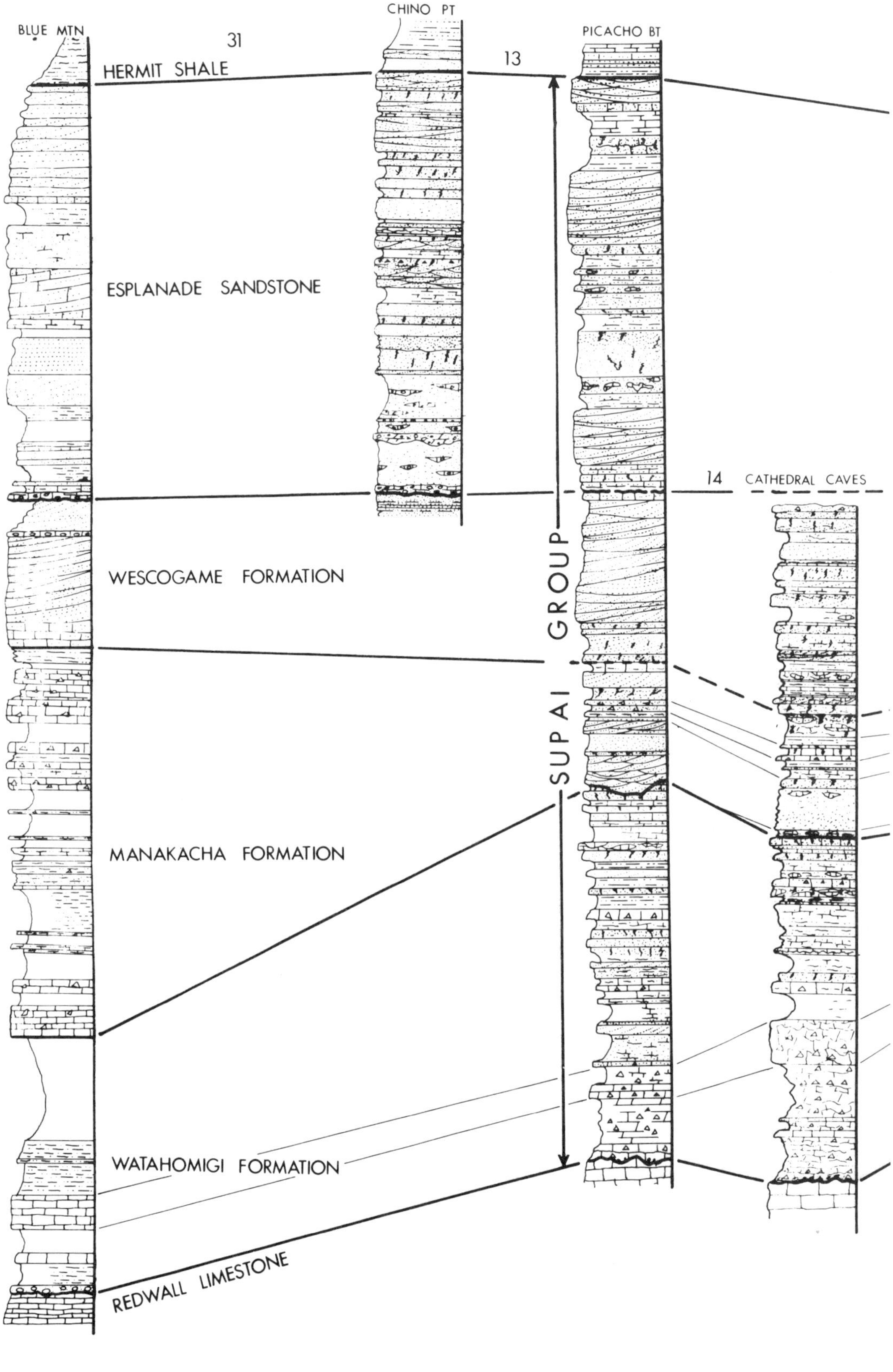

Figure 90. Columnar correlation of measured sections of Supai Group, central and western Mogollon Rim. Location provided by E. D. McKee (1978, written commun.).

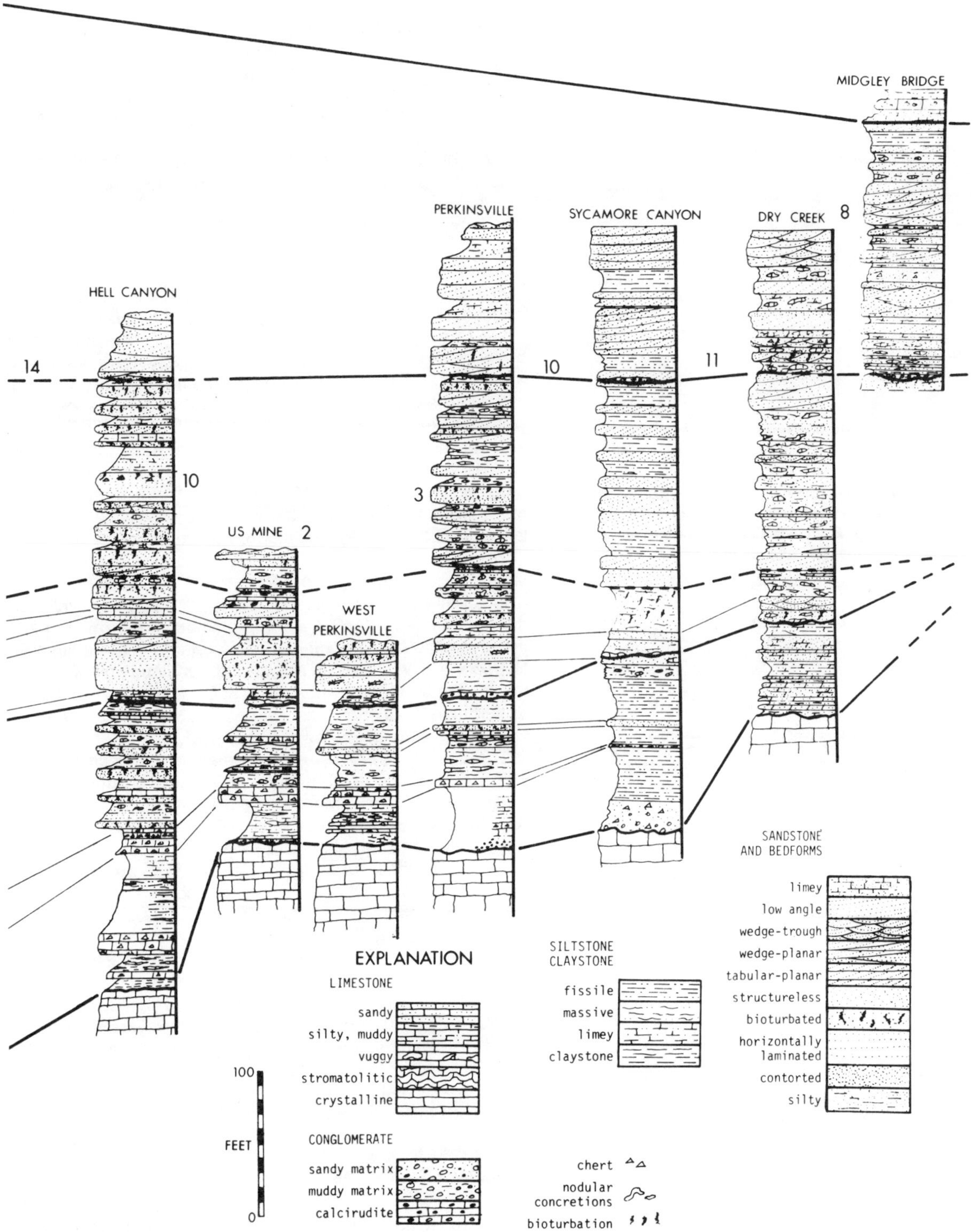

of sections shown on Figure 82. Distance between sections given in miles. Data for Blue Mountain Section

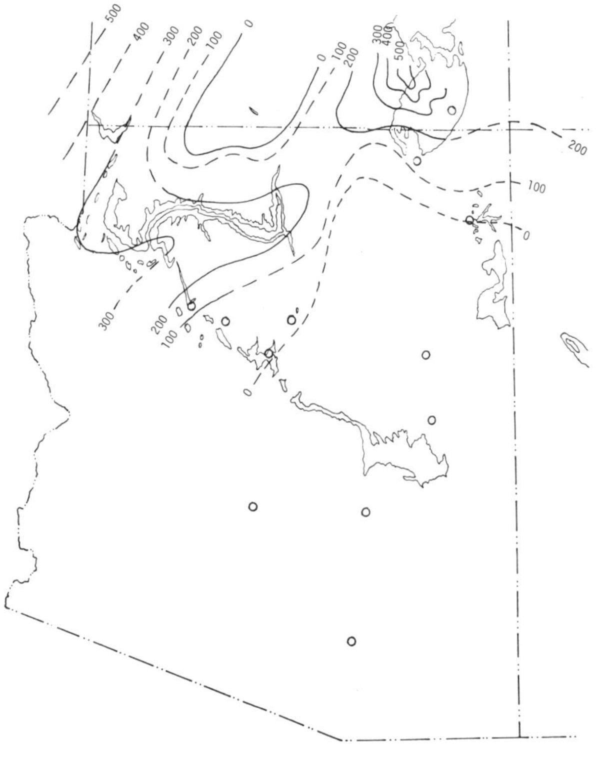

Figure 91. Isopach map of Manakacha Formation and related rocks, southern Colorado Plateau. Equivalent rocks in Four Corners area include the lower Hermosa Group (Pinkerton Trail Formation) and possibly parts of the Molas Formation. Isopachs given in feet.

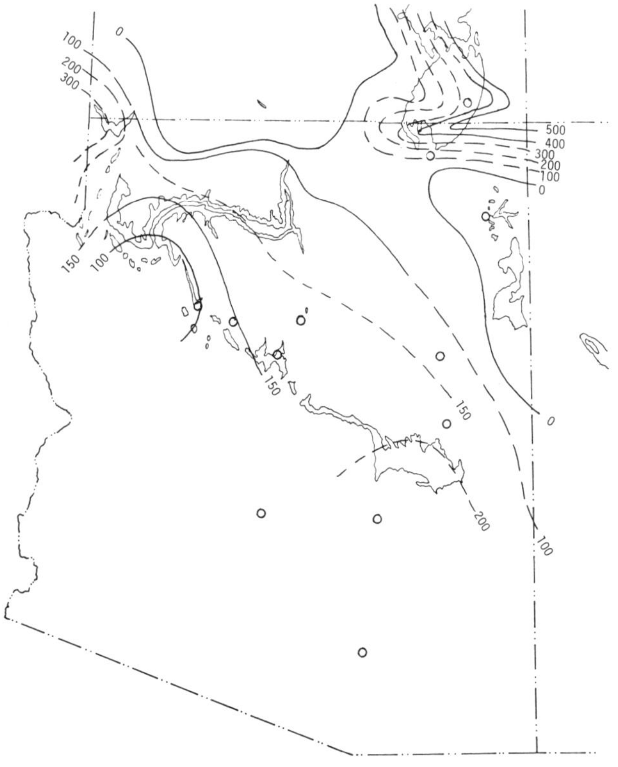

Figure 92. Isopach map of Wescogame Formation and related rocks, southern Colorado Plateau. Equivalent rocks include the upper Hermosa Group (Honaker Trail Formation) in Four Corners area and the Earp Formation in east-central Arizona. Isopachs given in feet.

does not readily stand out. However, on vertical cliffs and along streams 20 to 50 feet (6–16 m) of relief is generally apparent. In the Grand Canyon, McKee (1975) reported that at most localities the disconformity is overlain by conglomerate in the basal Esplanade Sandstone. In the west-central Mogollon Rim the conglomerate is absent in more places than it is present. The longest continuous exposure observed is in Oak Creek Canyon below Midgley Bridge, where the unconformity can be walked along the creek for over one-fourth of a mile. Local prominent pockets of conglomerate mark the boundary but in most places it is absent. A few yards south of Midgley Bridge on the west wall above an irrigation ditch, 30 or more feet (9 m) of relief are exposed on the rock wall. The lithologies immediately above and below the unconformity are similar and consist of dolomitic sandy siltstone. The precise amount of time missing at the boundary is difficult to determine because of a lack of fossils. In the eastern Mogollon Rim where some fossil control is available, Ross (1973) suggested that one or two fusulinid zones might be missing.

Several divisions can be correlated throughout the Mogollon Rim in the sandstone and siltstone sequence. Locally in the Perkinsville area the base of the Manakacha Formation is a slope-forming nodular and platy siltstone. This is succeeded by reddish-orange sandstone and silty sandstone that forms a cliff, broken cliff, or sharp ledge. This unit probably correlates to the Manakacha cliff unit of McKee (1975) in the Grand Canyon. This unit thickens uniformly from 17 feet (5 m) at Perkinsville to over 200 feet (60 m) in the western Grand Canyon. The amount of carbonate grains increases westward and at Iceberg Ridge, much of the Manakacha cliff would probably be more accurately described as limestone. The cliff unit is overlain by cyclic sandstone, siltstone, and limestone.

The portion of the sandstone and siltstone sequence assigned to the Wescogame Formation is not readily divisible. Sandstone is dominant lithologically and a strongly cyclic sequence is still evident. A fairly prominent cliff-forming, pink, cross-stratified sandstone can be correlated from Perkinsville to Sedona. The lower cliff unit and upper slope unit of McKee (1975) can be vaguely recognized at Picacho Butte but are not apparent in most sections to the east. Thickness of the Wescogame Formation is shown in Figure 92.

The age of the Manakacha Formation is Atokan (McKee, 1975). McKee also stated that it is possible, though not very likely, that some of the Manakacha is Desmoinesian. Because Atokan rocks are absent in eastern Mogollon Rim (Ross, 1973; Brew, 1965), the Manakacha does not correlate with the Naco Group (Horquilla Limestone) in that area. The area of deposition of the Manakacha Formation corresponds closely to that of the underlying Watahomigi Formation.

The Wescogame Formation is Virgilian in age (McKee, 1975). Virgilian-age rocks are much more widespread in the southern Colorado Plateau than rocks of the Lower and Middle Pennsylvanian. Wescogame is used in the Grand Canyon and west-central Mogollon Rim but equivalent rocks are present in southern Nevada (upper Callville Limestone), the Paradox Basin (Honaker Trail Formation), and the eastern Mogollon Rim (Earp Formation). A logical break in terminology along the Mogollon Rim would be between Oak and Fossil Creeks.

Esplanade Sandstone

The uppermost formation of the Supai Group, the Esplanade Sandstone, can be traced along a nearly continuous outcrop from Grand Canyon into the Mogollon Rim. The spectacular cliff and broad platform that the Esplanade typically forms in the Grand Canyon is not as prominent in the Mogollon Rim because of more complex gravel terraces, lava flows, and faulting in the latter region; however, the cliff and platform topography is locally displayed in several areas, notably the Sycamore Canyon area (fig. 93). It is likely that the somewhat more subdued outcrop of the Esplanade in the Mogollon Rim area has hindered previous recognition and correct correlation of this unit into the area.

The Esplanade Sandstone consists of thick cross-stratified sets of fine-grained, well-sorted, pale grayish-orange to light reddish-brown quartz sandstone. Thinner, slabby weathering, generally more silty units are interbedded with the more massive-weathering units and weather to notches in cliffs or broad irregular ledges. Where directly overlain by outcrops of the Hermit Shale, the Esplanade is typically stained reddish brown. Much of the Esplanade is calcareous and fossils, or weathered traces of fossils, can be seen in many thin sections. Fossil grains, some of which are strongly abraded, include foraminifers, bryozoans, brachiopods, mollusks, and, most abundant, echinoderms. Some of the smaller grains such as endothyrid foraminifers are likely reworked from Pennsylvanian rocks but the other fossil grains, particularly echinoderm fragments, show no sign of being part of previous rock and are likely contemporaneous with Esplanade deposition. Bioturbation of many types is present in the Esplanade and some burrows are silicified.

The Esplanade Sandstone thins progressively from northwest to southeast along the Mogollon Rim. The unit is approximately 300 feet (90 m) thick in the western Chino Valley and 170 feet (50 m) thick in the Sedona area. Typical Esplanade Sandstone is not present at Fossil Creek; the unit must either pinch out or change facies between Oak and Fossil Creeks. More work is needed to clarify this problem.

The lower boundary is an erosional unconformity, as has already been discussed. The upper boundary with the overlying Hermit Shale is not well exposed but no strong evidence for an unconformity, as can be seen in many places in Grand Canyon, was found along the Mogollon Rim. The boundary is placed at the top of the highest, light-colored, well-sorted, cross-stratified sandstone overlain by calcareous, poorly sorted, silty sandstone, conglomerate, or limey siltstone of the Hermit Shale.

The relationship of the Esplanade to the east and southeast of Sedona is not presently known. To the north and northeast the Esplanade correlates with the Cedar Mesa Sandstone and Halgaito Formation of the Monument Upwarp. According to Baars (1978, personal commun.), enough subsurface data is available to remove any doubt of the correlation between the Cedar Mesa and Esplanade. Westward and northward the Esplanade and Cedar Mesa grade into fossiliferous marine carbonate sequences. I have observed this transition between quartz sandstone and sandy carbonate in the Shivwits Plateau area of western Grand Canyon and it is being carefully documented by McKee (1978, personal commun.). In general the transition occurs farthest east at the base of the sequence and when followed to the north and west the carbonate gradually rises in section (fig. 94). The carbonate sequence is called the Pakoon Limestone in western Arizona and Nevada and the Elephant Canyon Formation in Utah.

No readily recognizable divisions are present in the Esplanade Sandstone in most of the Mogollon Rim. In adjacent sections similar sequences can be recognized but on a regional basis do not appear useful for correlation. In the Grand Canyon, McKee (1975) recognized two divisions at most localities. The lower division or Esplanade slope unit consists of sandstone, siltstone, and basal conglomerate. Westward it grades with the Pakoon Formation. It probably correlates with the Halgaito Formation of the southern Monument Upwarp. The upper unit or Esplanade cliff unit comprises mainly cross-stratified sandstone and correlates with the Cedar Mesa Sandstone.

Figure 93. Photograph of uppermost 20–30 feet (6–9 m) of Esplanade Sandstone near U.S. mine. Juniper-covered slope in background is underlain by Hermit Shale. Previous mapping in area (Lehner, 1958) included Esplanade in lower member of Supai Formation and Hermit in middle member of Supai Formation.

The age of the Esplanade Sandstone is Wolfcampian (McKee, 1975). The age is based on fossils found within the Pakoon Limestone. Baars (1962) concluded a similar age for the Cedar Mesa based on intertonguing with the fossiliferous Elephant Canyon Formation. Parts of the Esplanade and Cedar Mesa are fossiliferous in thin sections but to my knowledge, no serious attempt has been made to date the units using these fossils. At Sedona, the Esplanade Sandstone is 800 feet (245 m) below the Fort Apache Limestone which crops out near the middle of the Schnebly Hill formation. The Fort Apache yields an early Leonardian fauna (Winters, 1963; Gerrard, 1966).

Hermit Shale and Schnebly Hill formation

Only a brief account of these two units will be presented here. It should be remembered that the Hermit Shale and Schnebly Hill formation have been considered the middle and upper members, respectively, of the Supai Formation on the Mogollon Rim. They are extensively exposed in the Sycamore, Oak, and Fossil Creek areas and sporadically and incompletely exposed in western Verde and Chino Valleys.

The Hermit Shale can be traced through incomplete outcrops from Diamond Creek, through Chino Point, Picacho Butte, near Drake, and Perkinsville, and then by fairly continuous outcrops and subsurface control to Sedona and eventually eastward to Fort Apache where Finnell (1966) called it the Cibecue Member of the Supai Formation. Although similar to the Hermit in the Grand Canyon, the unit in the Mogollon Rim does differ. From Sedona eastward the Hermit contains abundant intraclastic limestone and siltstone. Near Sedona 18 separate conglomerates were measured in one stratigraphic section. Although this is extreme, most other sections in the central Mogollon Rim area have at least four or five conglomerate horizons. Also, the Hermit along the Mogollon Rim is more calcareous than in the Grand Canyon. Local, fairly pure carbonate beds are abundant from Sedona eastward. From the eastern Mogollon Rim to Sycamore Canyon the Hermit maintains a thickness of about 300 feet (90 m). It thins to 90 feet (27 m) at Chino Point and then thickens drastically to over 800 feet (240 m) in the Shivwits Plateau area.

The Hermit Shale is abruptly succeeded by the Schnebly Hill formation. The Schnebly Hill forms much of the spectacular red rock scenery in the Sedona area. The unit comprises cross-stratified sandstone, ripple-laminated silty sandstone, and dolomite in the central and western Mogollon Rim. Eastward a more basinal facies contains siltstone, gypsum, limestone, and mudstone as well. The Schnebly Hill thickens eastward and northward from western Chino Valley. The unit is absent at Chino Point; however, 91 feet (28 m) are recognized 14 miles (22 km) east at Picacho Butte. It is approximately 800 feet (240 m) thick in the central Mogollon Rim.

The Schnebly Hill formation grades laterally and vertically into the Coconino Sandstone. Spectacular intertonguing can be seen throughout the Sedona area. At the base of the upper third of the unit a prominent limestone-dolomite marker, the Fort Apache Limestone, is present along the central and eastern Mogollon Rim. From a zero edge a few miles west of Sedona, the Fort Apache thickens to 100 feet (30 m) in the Fort Apache area. Although many workers previously considered the Fort Apache to be a member of the Supai Formation, the restriction of the Supai on the Mogollon Rim in this paper leaves the Fort Apache without formal designation at this time. A later publication in preparation will assign it member status in the Schnebly Hill formation.

The Schnebly Hill is considered latest Wolfcampian and earliest Leonardian in age, based on an early Leonardian fauna in the Fort Apache Limestone (Winters, 1963; Gerrard, 1965).

DEPOSITIONAL HISTORY

The Supai Group and related rocks of the Mogollon Rim form part of a complex continental-marine sequence that was deposited during the Pennsylvanian and Early Permian. Numerous previous workers have loosely assigned Supai deposition to a deltaic environment. Both stratigraphic and sedimentological data gathered in this study contradict these earlier findings. Vertical sequences typical of deltaic environments are not abundant and paleogeography and basin analysis do not support major deltaic episodes of deposition. Detailed reconstructions of environments and sequences of transgression and regression are not possible at this time but preliminary conclusions can be presented.

Very little of the Supai clastic material came from the Uncompahgre, Zuni, or Defiance Upwarps to the east, although they had local influence. Strong evidence based on facies control and cross-stratification readings demonstrates that the sand and coarse silt was derived from the north and northwest. McKee (1969; 1978, personal commun.) has evidence in the Grand Canyon supportive of this conclusion. Thus Supai clastics were transported by longshore processes from northern Utah, Wyoming, and adjacent areas southward through central Utah across and around the west side of the Emery

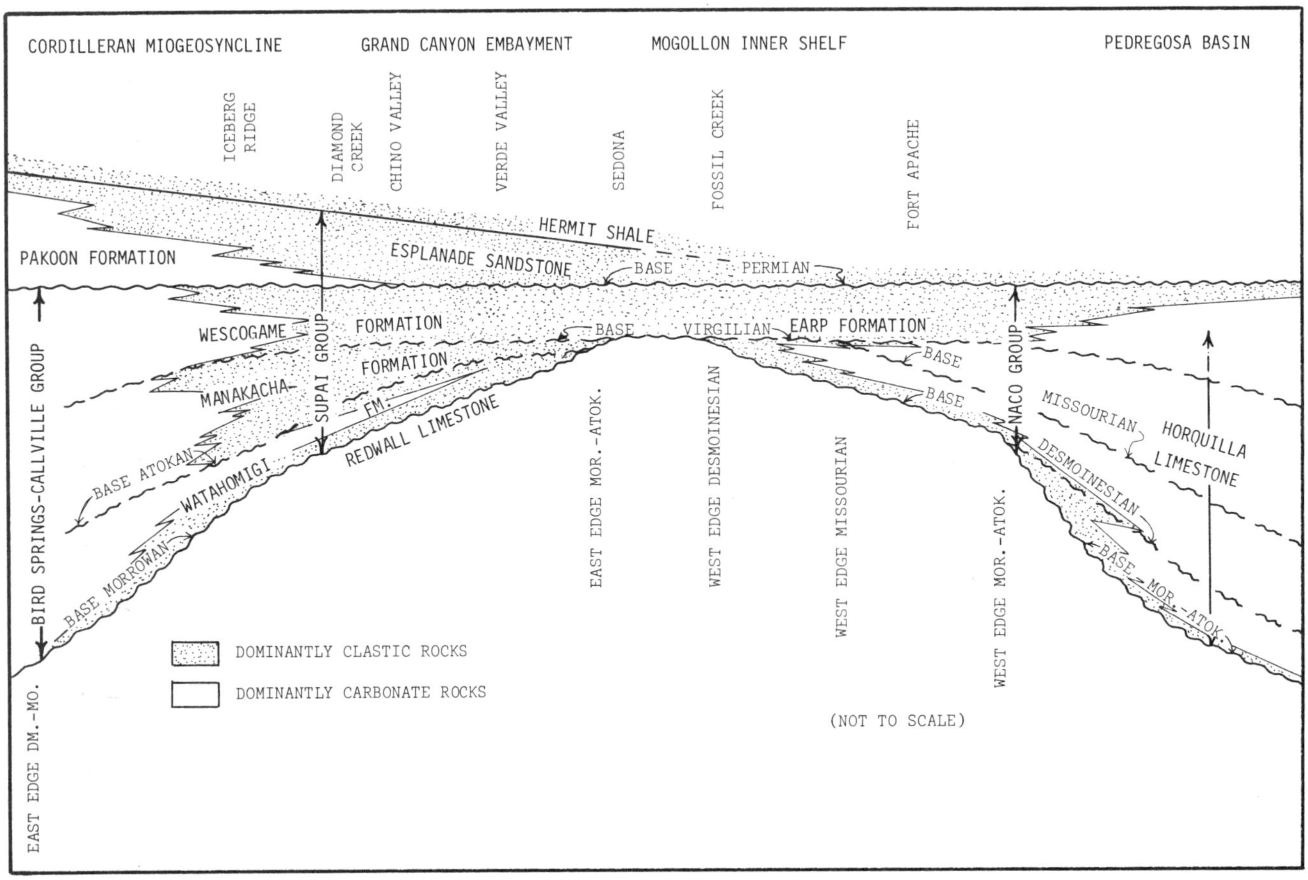

Figure 94. Time-rock stratigraphic chart of Supai Group and related rocks from southern Nevada to southeastern Arizona. Note that Supai and Naco Groups do not intertongue; of the Pennsylvanian rocks, only Virgilian strata cross structural high near Sedona. For more detail and true scale, compare with Figures 83, 86, and 90.

Uplift (Piute Platform) into northern Arizona. A number of medium-grained clastic units to the north, including the Weber, Tensleep, and Quadrant Sandstones, of Middle and Late Pennsylvanian and Early Permian age, document abundant sand in this direction. Coarse-clastic debris being shed from the Uncompahgre Uplift was trapped on the edge of the Paradox Basin and did not reach northern Arizona. The Defiance area, although positive, was not an important source of sediment at this time.

The Grand Canyon Embayment and Mogollon Shelf areas were the site of numerous marine transgressions and regressions during Pennsylvanian and Early Permian times. The major seaway was to the northwest in the Cordilleran Miogeosyncline. During times when clastic debris was cut off by local or regional processes, carbonate deposition took place. In the Mogollon Rim, significant carbonate deposition occurred in the Watahomigi Formation. At Cathedral Caves, 40–50 percent of the unit is carbonate. The percentage decreases eastward to less than 15% at Perkinsville. Marine fossils are present at Cathedral Caves and Picacho Butte. Most other original sedimentary textures are recrystallized and few direct clues to depositional history are available. Some carbonate units have wispy and laminar bedding suggestive of stromatolites. Probably most carbonates in the Watahomigi Formation were deposited in nearshore marine and shoreline environments.

Clastic deposition in the Supai Group is complex and represents several different environments. Much of the medium and large scale cross-stratified units were likely deposited as bar, shoal, and beach deposits on tidally influenced shelves. Evidence includes broad, sheet geometry of most sand bodies, clean, well-sorted, relatively mature sandstone, local herringbone pattern tabular-planar cross-stratification, local possible flaser bedding, abundant carbonate grains including abraded marine fossils, intense bioturbation in many units, and westward gradation into known marine carbonate and sandstone sequences. The upper Watahomigi, Manakacha, and lower Wescogame Formations display the most cyclic sequences. Presumably the different components of the cycles represent different physical, chemical, and biologic parameters of the transgressing and regressing shallow seas. These cyclic deposits need additional analysis before they can be completely interpreted.

Thicker sand bodies in the Manakacha cliff unit, the upper Wescogame Formation, and much of the Esplanade Sandstone probably were deposited in environments similar to those of the above sandstone units but represent more stable, widespread conditions. The sand was probably distributed, deposited, and reworked by shallow marine currents. Larger scale cross-stratification and less bioturbation in some of these units suggest higher energy environments that reworked sand bodies and destroyed bioturbation.

Following deposition of the Esplanade Sandstone by nearshore processes, the southern Colorado Plateau was inundated by fine-grained red clastics of the Hermit Shale. The clastics were derived from the east and northeast. Much of the Hermit Shale of the Grand Canyon has been considered fluvial although a modern, comprehensive study is necessary to document this. A higher percentage of limestone and limestone-pebble conglomerate along the Mogollon Rim suggests somewhat different conditions from the Grand Canyon. Numerous stromatolitic features including oncolites suggest shoreline deposition. Stromatolitic carbonate was then ripped up during storms or reworked by fluvial processes during lower stands of the sea.

The Schnebly Hill formation was deposited on the edge of and within the Holbrook Basin. Preliminary work suggests deposition in a complex sebkha and shallow marine system. Supratidal, intertidal, and subtidal conditions alternated with aeolian processes as the Coconino dunes prograded southward into the early Leonardian sea.

CONCLUSIONS

1. The previous Supai Formation of the Mogollon Rim included rocks of Pennsylvanian and Permian age. Present stratigraphic work indicates only the lower portion of the Supai Formation correlates with the type Supai Group. The four formations of the group can be recognized as far east as Dry Creek near Sedona. The so-called middle and upper members of the Supai Formation are younger than the type Supai. The middle unit is the Hermit Shale, and the upper unit is assigned to a new unit informally called the Schnebly Hill formation.

2. Most of the newly defined Supai Group along the western Mogollon Rim is of different age than the Naco Group of the eastern Mogollon Rim country (fig. 94). Therefore there is no true intertonguing between the two units. The Watahomigi and Manakacha Formations of the Supai Group are Morrowan and Atokan in age and the Horquilla Limestone of the Naco Group is Desmoinesian and Missourian in age. The Wescogame Formation of the Supai Group and the Earp Formation of the Naco Group are part of the same continuous unit; at Sedona and westward Wescogame is used and at Fossil Creek and eastward Earp is used. The Esplanade Sandstone may not have been deposited east of the Sedona area.

3. As defined in this paper, the Supai Group of the Mogollon Rim consists of four formations. The basal Watahomigi Formation comprises basal conglomerate, carbonate, and cyclic red clastics. The Manakacha Formation includes cross-stratified red sandstone and complex cyclic clastics and carbonates. The Wescogame Formation consists of reddish-orange cyclic sandstone and siltstone. The Permian Esplanade Sandstone comprises cross-stratified, mostly light-colored, cliff-forming sandstone. Although the boundaries between the Watahomigi, Manakacha and Wescogame Formations are locally difficult to pick, each unit has characteristics that serve to separate it from adjacent units.

4. The Supai Group was deposited in complex marine and paralic environments. The overlying Hermit Shale was deposited in a fluvial-tidal complex, and the Schnebly Hill formation represents transitional environments between the Leonardian sea to the south and Coconino dunes encroaching from the northwest.

REFERENCES CITED

Baars, D. L., 1962, Permian system of Colorado Plateau: American Association of Petroleum Geologists Bulletin, v. 46, p. 149–218.

Blakey, R. C., 1978, Pennsylvanian and Lower Permian paleogeography and paleotectonics of the southern Colorado Plateau: Geological Society of America Abstracts with Programs, v. 10, no. 3, p. 96.

Brew, D. C., 1965, The Naco Formation (Pennsylvanian) in central Arizona: Cornell University, unpublished Ph.D. dissertation, 201 p.

Finnell, T. L., 1966, Geologic map of the Cibecue Quadrangle, Navajo County, Arizona: United States Geological Survey Map GO–545.

Gerrard, T. A., 1966, Environmental studies of the Fort Apache Member, Supai Formation, east-central Arizona: American Association of Petroleum Geologists Bulletin, v. 50, p. 2434–2463.

Gordon, Mackenzie, Jr., and McKee, E. D., 1978, Significance of the invertebrate fauna of the Watahomigi Formation (Supai Group): Geological Society of American Abstracts with Programs, v. 10, no.3, p. 107.

Huddle, J. W., and Dobrovolny, E., 1945, Late Paleozoic stratigraphy of central and northeastern Arizona: United States Geological Survey Oil and Gas Investigations, Preliminary Chart 10.

Hughes, P. W., 1952, Stratigraphy of the Supai Formation, Chino Valley, Yavapai County, Arizona: American Association of Petroleum Geologists Bulletin, v. 36, p. 635–657.

Jackson, R. L., 1952, Stratigraphic relations of the Supai Formation of central Arizona: Plateau, v. 24, p. 84–91.

Krieger, M. H., 1965, Geology of the Prescott and Paulden Quadrangle, Arizona: United States Geological Survey Professional Paper 467, 127 p.

Lane, C. L., 1977, Pennsylvanian-Permian stratigraphy of west-central Arizona: Northern Arizona University, unpublished M.S. thesis, 120 p.

Lehner, R. E., 1958, Geology of the Clarkdale Quadrangle, Arizona: United States Geological Survey Bulletin 1021–N, p. 511–592.

McKee, E. D., 1945, Oak Creek Canyon: Plateau, v. 18, p. 25–32.

———1975, The Supai Group, Subdivision and nomenclature: United States Geological Survey Bulletin 1395–J, 11 p.

———, and Gutschick, R. C., 1969, History of the Redwall Limestone of northern Arizona: Geological Society of America Memoir 114.

Peirce, H. W., Jones, N., and Rogers, R., 1977, A survey of uranium favorability of Paleozoic rocks in the Mogollon Rim and Slope region, east central Arizona: Arizona Bureau of Geology and Mineral Technology Circular 19, 60 p.

Ross, C. A., 1973, Pennsylvanian and Early Permian depositional history, southeastern Arizona: American Association of Petroleum Geologists Bulletin, v. 57, p. 887–912.

Winters, S. S., 1963, Supai Formation (Permian) of eastern Arizona: Geological Society of America Memoir 89, 99 p.

CHARACTERISTICS OF THE SUPAI GROUP IN GRAND CANYON, ARIZONA

Edwin D. McKee[1]

Abstract — In the Grand Canyon region the Supai Group, as currently defined, consists of four formations, three of them separated by unconformities of erosion. These formations have been examined from the standpoints of paleogeography, rock distribution and thickness, faunal distribution, clay minerals, texture and structure of sandstones, and carbonate facies.

The Watahomigi or oldest formation is interpreted as having formed primarily in a low-energy marine environment. The overlying Manakacha Formation is also considered to be mostly marine, but high energy. The third oldest, or Wescogame, is believed to be fluvial, possibly braided stream, and high energy. The youngest, or Esplanade Sandstone, is difficult to interpret for it shows some features of continental and some of marine environment. Possibly it is estuarine.

NOMENCLATURE AND AGE

A prominent series of red-bed strata, ranging from 700 to more than 1,000 feet (215–300 m) thick, occurs throughout Grand Canyon in its upper half and is referred to as the Supai Group. The name "Supai" was first applied to these rocks by Darton (1910, p. 25), because "They are very conspicuously exposed near Supai village" (fig. 95) in the bottom of Havasu (Cataract) Canyon, a tributary of Grand Canyon on its south side (fig. 96).

As originally defined, the Supai was considered a single formation, primarily because it consisted mostly of red sandstones and shaly mudstones throughout, and because neither marker beds nor evidence of unconformities had been observed to suggest natural subdivisions. Subsequent detailed studies, however, have shown that the Supai is composed of four well-defined, mappable units (fig. 97), separated by erosion surfaces, key conglomerate beds, or both (McKee, 1975). Furthermore, the subdivisions differ considerably one from another, in details of composition, geologic age, environment of deposition, faunal distribution and content, paleogeography, and other attributes.

Because of the distinctive characteristics recognized in subdivisions of the Supai, its rank has been elevated to that of group and the component parts are given the status of formations (McKee, 1975, p. J2). Names applied to the formations are, in ascending order: Watahomigi, Manakacha and Wescogame Formations and Esplanade Sandstone. The first three names were derived from physiographic features in Havasu Canyon (fig. 95), and their type sections are in that area. The uppermost formation, the Esplanade Sandstone, is named from the great sandstone bench, known as the Esplanade Platform, that this rock forms throughout central and western Grand Canyon.

Ages of the four formations[2] composing the Supai Group were long in doubt because fossil evidence was sparse and largely nondiagnostic. With the discovery of numerous faunal localities and the gradual recognition and extension of stratigraphic units during the past 30 years, however, the age of

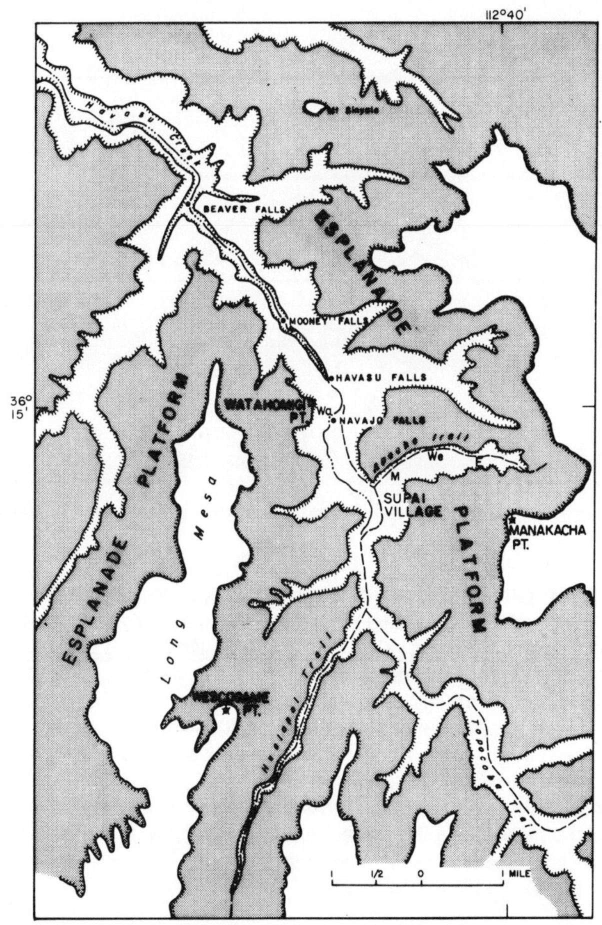

Figure 95. Havasu Canyon, Arizona, showing locations of type sections. Wa – Watahomigi Formation; M – Manakacha Formation; We – Wescogame Formation; E – Esplanade Sandstone.

---

[1] U.S. Geological Survey, Denver, Colorado 80225.

[2] Faunal determinations were made by the following: small foraminifers—L. G. Henbest (retired), and Bernard Mamet (University of Montreal); fusulinids—Raymond C. Douglass, corals—W. J. Sando, brachiopods—Mackenzie Gordon, Jr., gastropods—E. L. Yochelson, pelecypods—John Pojeta, Jr., all of the U.S. Geological Survey.

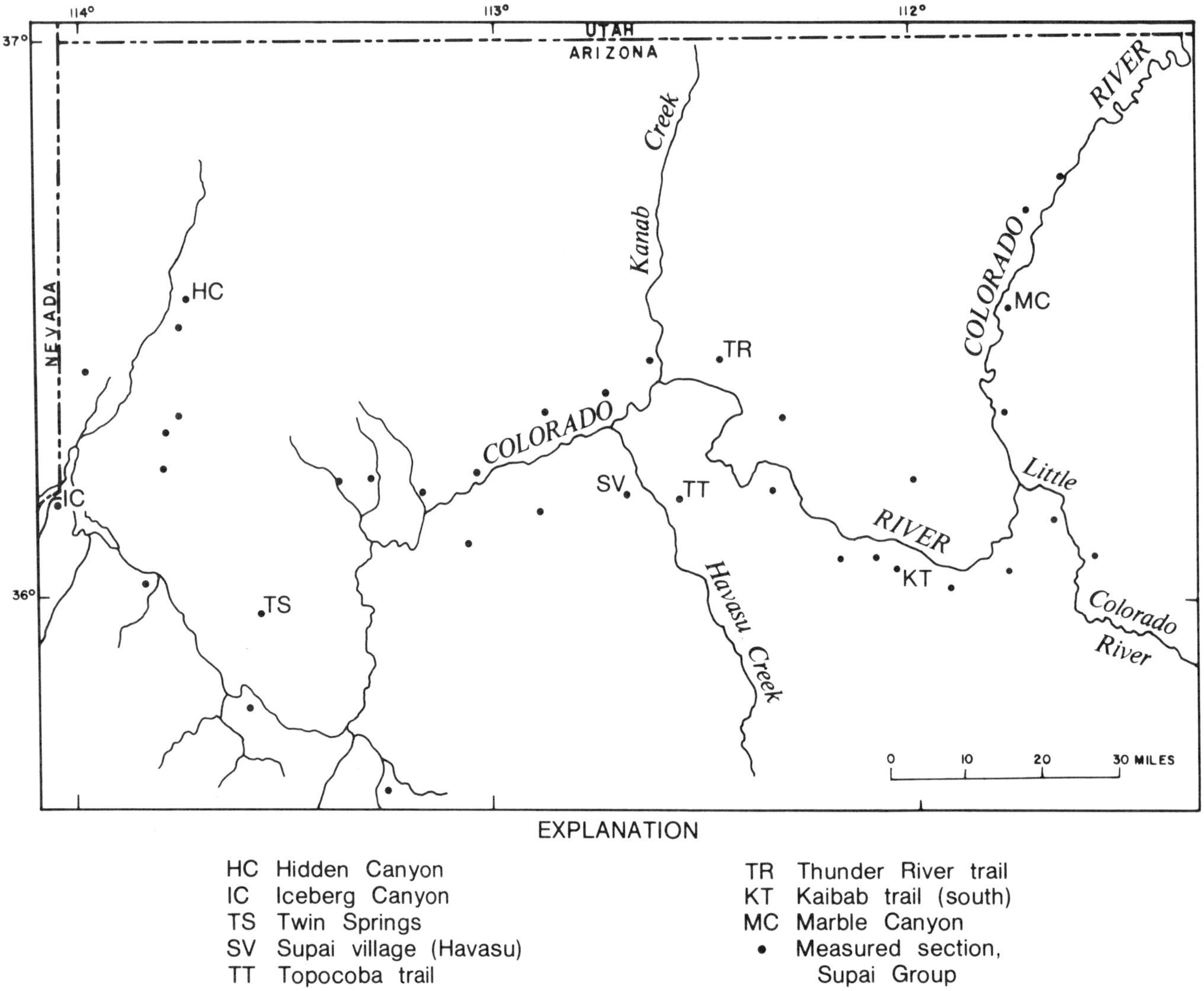

Figure 96. Grand Canyon, Arizona, and locations of measured sections of Supai Group.

each formation has been established. Largely on the basis of brachiopods, the lower two units of the Watahomigi are considered to be of Morrowan age. Based on fusulinids and brachiopods, the upper unit of the Watahomigi Formation has been determined to be Atokan. Age determinations for the upper three formations are based mostly on fusulinid occurrences, although some supporting evidence has been obtained from smaller foraminifers, corals, and other fossil groups. The ages as currently assigned to the formations are: Manakacha (lower or cliff unit), Atokan; Manakacha (upper or slope unit), Atokan and Des Moinesian(?); Wescogame, Virgilian; Esplanade Sandstone, Wolfcampian.

DISTRIBUTION AND THICKNESS

The structural framework that controlled the distribution and thickness of deposits in the Grand Canyon region during Pennsylvanian and Early Permian time included three consistently negative elements, two intermittently negative passages or sags, and three consistently positive elements (fig. 98). The three controlling negative elements were the Cordilleran trough to the west and northwest, the Sonoran trough to the southeast, and the Paradox basin to the northeast. Temporary negative passages were the Arizona sag in central Arizona and the Black Mesa sag in northeastern Arizona. Positive elements were the Defiance-Zuni element to the east near the Arizona-New Mexico border, the Piute element to the north in Utah, and a questionable Ensenada Land to the southwest. Isopach maps, prepared for each formation of the Supai Group (fig. 99A–D) show major differences in geometry and paleogeography, and the effects of the positive and negative elements on their development.

Strata of the Watahomigi were deposited in a semicircular embayment which extended eastward from the Cordilleran geosyncline of southern Nevada. The embayment was landlocked on three sides and the deposits in it had a maximum thickness of between 200 and 250 feet (60–76 m), located near the middle.

The embayment during the time of deposition of the Manakacha Formation was similar to that of the underlying Watahomigi in general shape and relations to the western seaway. It differed radically, however, in having a northeastward connection with a relatively deep basin in southeastern Utah. Greatest thicknesses of the Manakacha, in excess of 300 feet

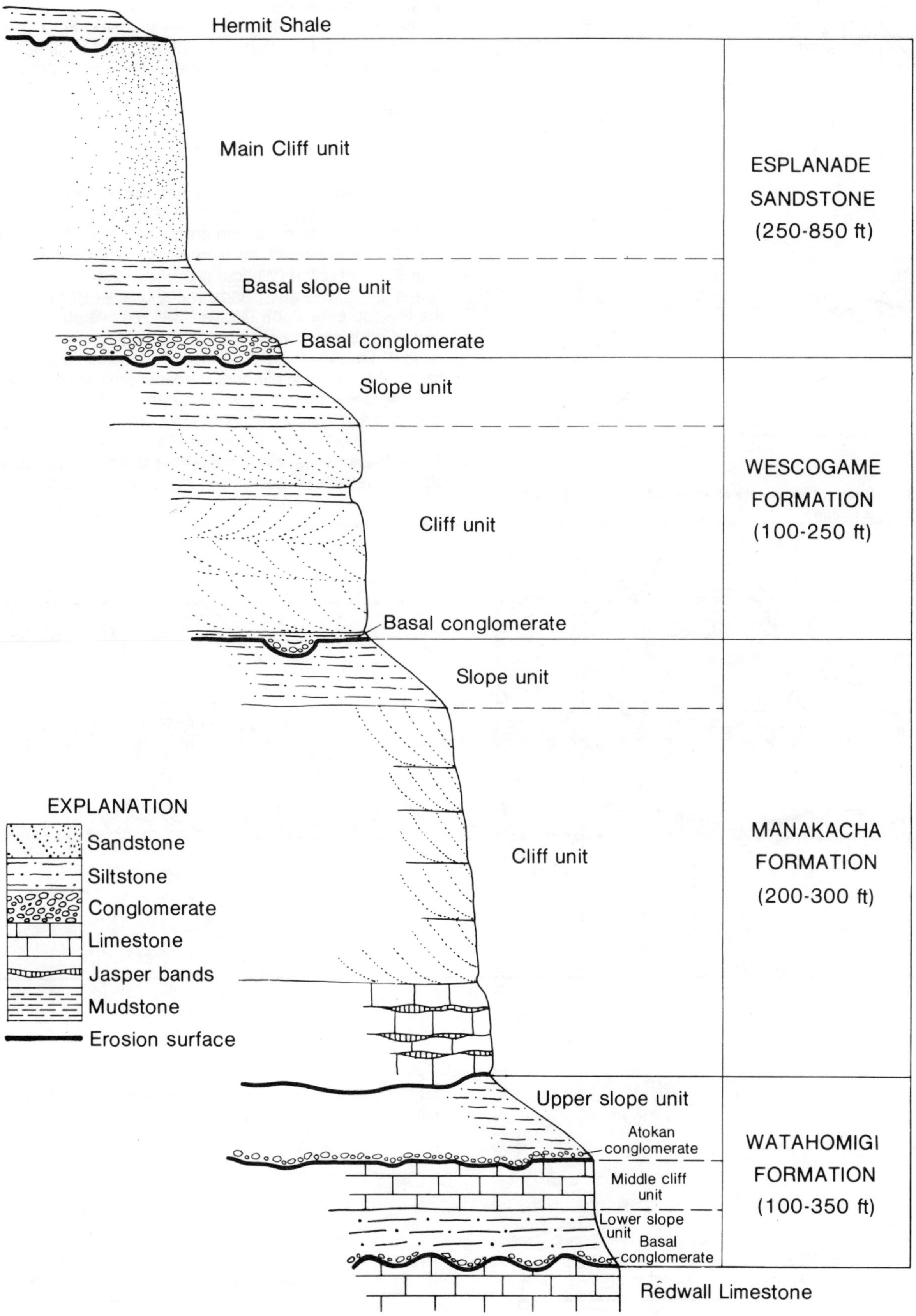

Figure 97. Generalized section of Supai Group showing formations, principal rock units, unconformities, and key conglomerate beds.

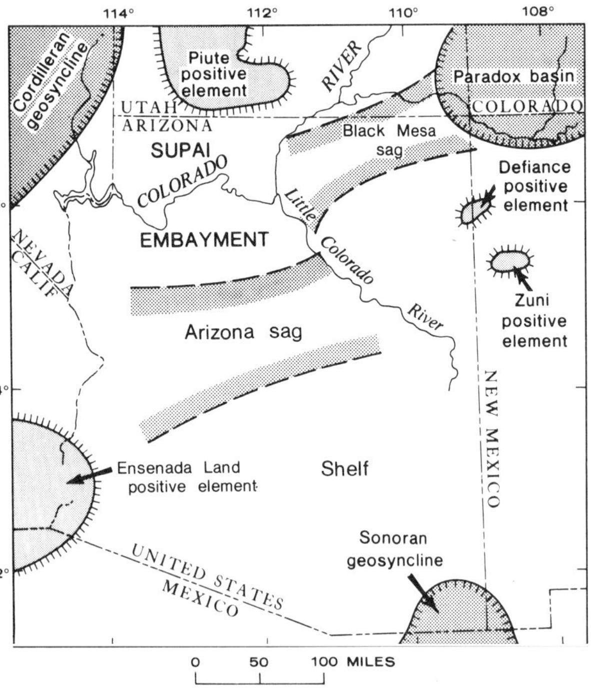

Figure 98. Structural framework of Arizona during depositional history of Supai Group.

(90 m), are in the central Grand Canyon area. Land areas were located to the north, east, and southeast.

The depositional area of the Wescogame Formation was more extensive than that of the underlying Manakacha, for it not only connected with a basin to the northeast, but also seems to haye extended across a broad area to the southeast and was continuous with deposits of the Sonoran geosyncline in southern Arizona and Mexico. When the Wescogame was formed, during Virgilian time, the only positive elements remaining as land masses in the Grand Canyon region were to the north and the east. A maximum thickness of the Wescogame of slightly less than 250 feet (76 m) is recorded from the southeastern Grand Canyon area, and resulted in a thin, tabular sheet of sediment, the thinnest in the Supai Group.

In Early Permian (Wolfcampian) time the depositional surface of the Esplanade Sandstone and its western correlative, the Pakoon Limestone (McNair, 1951), covered nearly every part of northern Arizona so that the only remnants of land to remain above the general surface of deposition were in the Defiance-Zuni positive element. Sediment thickness of 400 feet (120 m) accumulated in various parts of the Grand Canyon area and a maximum of more than 1,000 feet (300 m) was deposited a short distance northwest of the Grand Canyon area. These thicknesses testify to the magnitude of the depositional blanket that was spread across the region during this time.

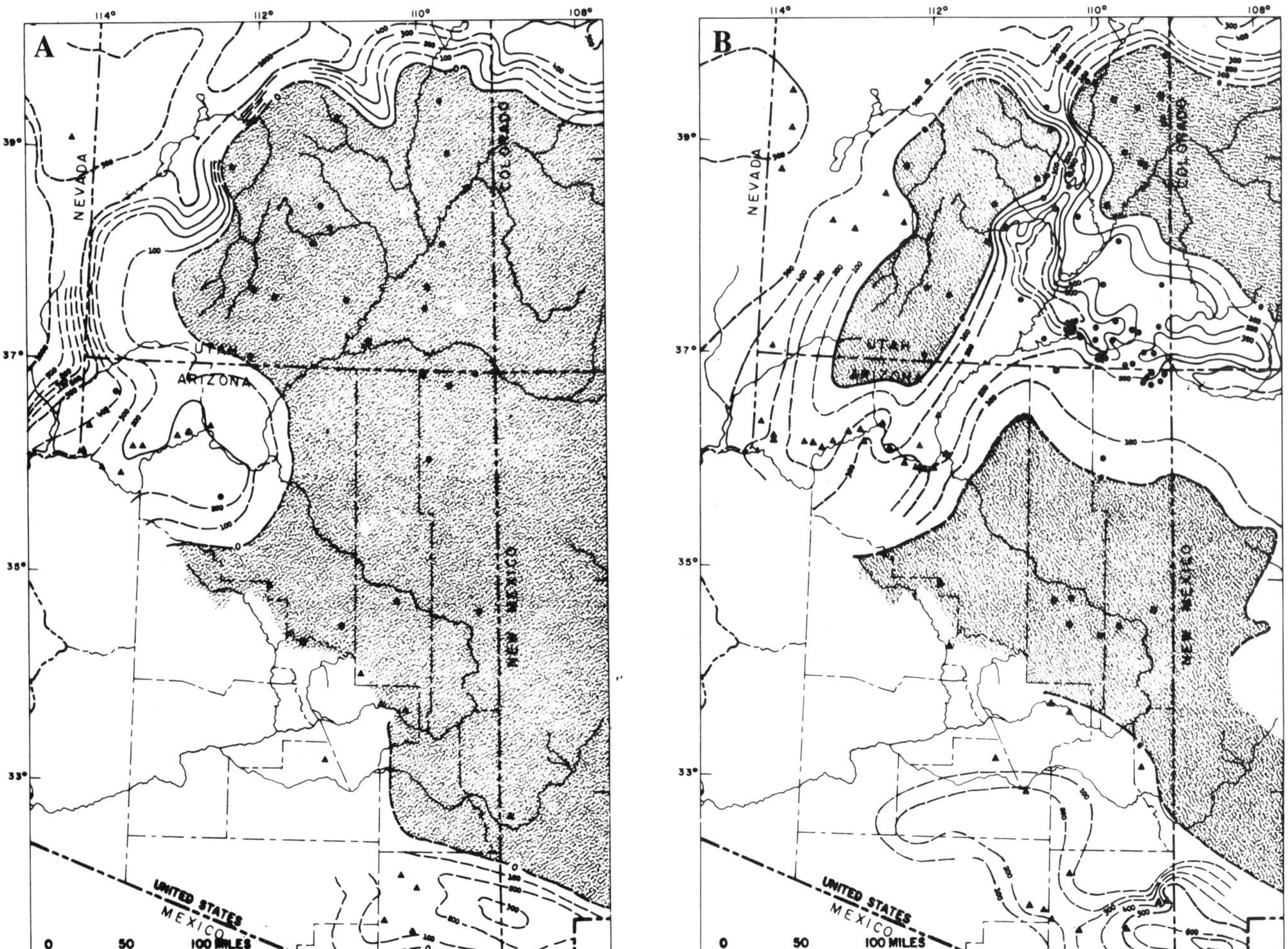

Figure 99. Isopach-paleogeographic maps of Arizona during time of deposition of each formation in Supai Group. A – Morrowan; B – Atokan; C – Virgilian; and D – Wolfcampian. Shaded area represents land. Isopachs show thickness of continental and marine sediments combined; dashed in areas of poor control. Isopach intervals, 100 feet and 500 feet (30–150 m). Data from outcrops (solid triangles) and wells (solid circles) represent thickness control points.

ANALYSIS OF LITHOLOGY

The Watahomigi (fig. 100B, C), oldest formation of the Supai Group, is composed largely of carbonate rock and mudstone (both silt and argillaceous mud), but relatively little sandstone. The distribution of these lithologies within the embayment is indicated by the following calculations:

	Eastern Grand Canyon	*Western Grand Canyon*
Carbonate rock	15–35%	60–90%
Sandstone	15%	Absent
Mudstone (includes siltstone)	50–85%	10–40%

The Manakacha Formation (fig. 100A, C) overlying the Watahomigi differs greatly from it in containing a considerable amount of sandstone throughout the area. It likewise includes much siltstone and argillaceous mudstone in eastern Grand Canyon but very little in the western part. Distribution patterns of the principal components in this formation are as follows:

	Eastern Grand Canyon	*Western Grand Canyon*
Carbonate rock	5–20%	35–90%
Sandstone	20–50%	10–30%
Mudstone (includes siltstone)	30–60%	10%

By far the most important rock type in the Wescogame is sandstone (fig. 100D, F), except in extreme western Grand Canyon where limestone (fig. 100E) is dominant. Mudstone is relatively inconspicuous in most of the area and carbonate rocks are sparse or absent in many eastern localities. Calculations of rock distribution for the Grand Canyon area are:

	Eastern Grand Canyon	*Western Grand Canyon*
Carbonate rock	0–30%	50–75%
Sandstone	70–90%	10–50%
Mudstone (includes siltstone)	20–35%	0–10%

The uppermost formation of the Supai Group, the Esplanade Sandstone (fig. 101A, B), is largely composed, throughout the area, of very fine-grained, cross-stratified sandstone, but includes some units of siltstone and mudstone; in the northwest, there is a facies composed of gypsum and gypsiferous mudstone. In extreme western Grand Canyon, a stratigraphic equivalent of the lower part of the Esplanade, referred to as the Pakoon Limestone, interfingers with it.

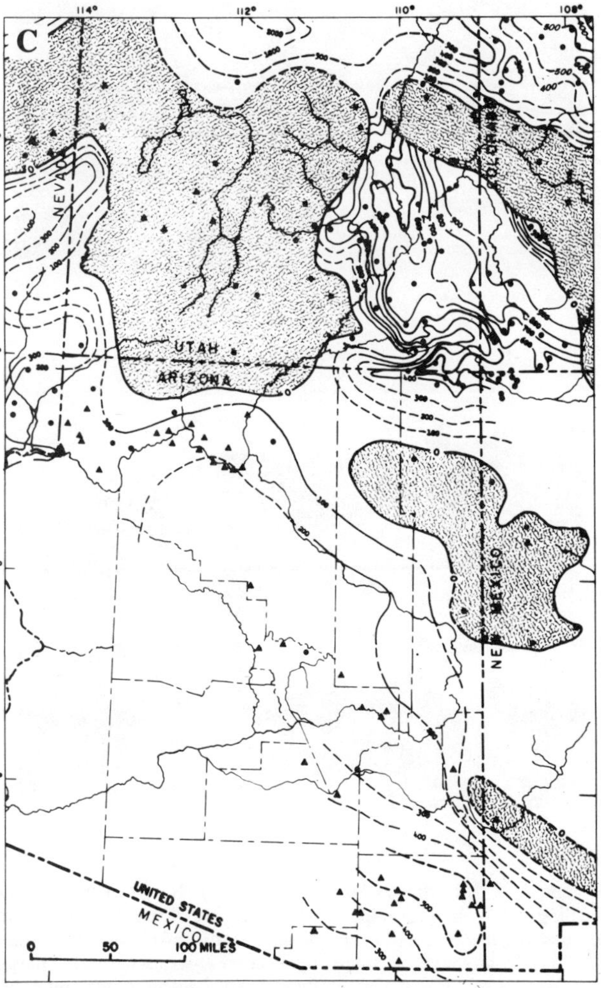

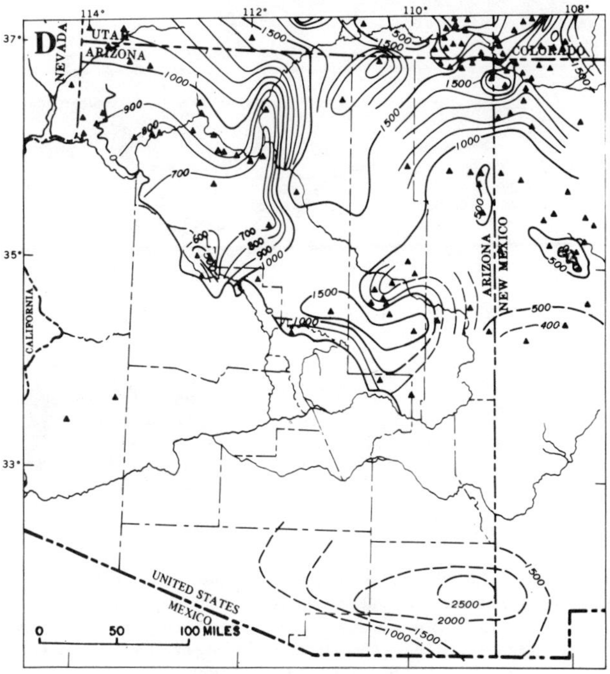

Figure 99 (Continued). For explanation see facing page.

Figure 100. Lithology and structure of Supai Group in Grand Canyon, Arizona. A – Manakacha Formation, lower cliff unit of thick-bedded, calcareous sandstone and sandy limestone. Wall of Havasu Canyon near Supai village; B – stratigraphic section from base of Supai Group to top of Kaibab Limestone. Shivwits Plateau near Twin Springs; C – Supai Group and overlying Hermit Shale near type section at Supai village; D – fifteen-foot set of tabular-planar cross-strata in Wescogame Formation. Topocoba trail, Grand Canyon; E – tabular-planar cross-strata in sandy limestone of Wescogame Formation. Iceberg Canyon, west of Grand Canyon; F – large-scale tabular-planar cross-strata at top of sandstone cliff unit of Wescogame Formation (We), Esplanade Sandstone (E) beyond. Thunder River trail, Grand Canyon. Symbols on photographs: Wa – Watamomigi Formation; M – Manakacha Formation; We – Wescogame Formation; E – Esplanade Sandstone; H – Hermit Shale; C – Coconino Sandstone; T – Toroweap Formation; K – Kaibab Limestone.

A

B

Figure 101. Lithologic features of Esplanade Sandstone; Grand Canyon, Arizona. A – Massive sandstone forming rim rock that overlies horizontal thin beds of sandstone. Thunder River trail; B – Large-scale trough cross-strata forming festoons. Hidden Canyon. Photograph by E. H. McKee.

FAUNAL DISTRIBUTION

Marine fossils of the Watahomigi comprise an assemblage containing many individuals, but relatively few genera and/or species. They extend eastward as far as Havasu Canyon in east-central Grand Canyon and include brachiopods, pelecypods, gastropods, corals, trilobites, bryozoans, and others. In eastern Grand Canyon, although well-preserved invertebrate fossils have not been found, most groups are represented in scattered accumulations of bioclasts in limestones. Most common among these are pelmatozoans, foraminifers, bryozoans, and bivalves.

In the Manakacha, invertebrate marine fossils seem to be facies-controlled and occur largely in the westernmost Grand Canyon area where they are restricted to the limestones. Bioclasts formed largely of pelmatozoans, calcisphaerids, and algae are found at various localities in central and eastern Grand Canyon, probably reflecting the development of a continuous passage of the sea from the Nevada area on the west to the San Juan basin on the northeast. Furthermore, fusulinids of southern Nevada are represented by three genera in western Grand Canyon and two in the eastern area.

CLAY MINERALS[3]

Clay minerals constitute a small, but significant part of each formation in the Supai Group and, in many places, seem to be good indicators of the various environments of deposition. They occur in association with sandstones, carbonate rocks, and siltstones or mudstones. Commonly they fill small pockets in massive beds, but they also occur as lenses or occupy thin parting planes between thick strata of sandstone or limestone. Six principal clay minerals have been recognized among approximately 800 samples that were examined from the four formations. They are illite, corrensite, kaolinite, chlorite, montmorillonite, and mixed-layer (illite-smectite) clays.

In the Watahomigi Formation of western Grand Canyon, illite, chlorite, and corrensite are commonly associated with carbonate rocks and are interpreted to be of marine origin. Kaolinite, which is common in eastern Grand Canyon, is considered to be of continental origin. Some illite and corrensite, also present in that area, may indicate the result of occasional marine incursions. This interpretation can be substantiated by the presence of much carbonate mudstone (micrite) and by a general absence of sandstone.

The Wescogame contains a marine fauna consisting of fusulinids, pelecypods, gastropods, and pelmatozoans, largely restricted to carbonate facies in western Grand Canyon. No brachiopods have been recorded. Apparently the medium- to large-scale cross-strata of the sandstone that constitutes most of the formation indicate high-energy conditions and an inhospitable environment for most forms of life. In eastern Grand Canyon, the well-preserved footprints of amphibians or reptiles occur on foreset beds at a number of localities.

The Pakoon Limestone in westernmost Grand Canyon intertongues with the lower part of the Esplanade Sandstone to the east and contains abundant fossils, especially fusulinids, corals, bivalves, and pelmatozoans. For the most part, these fossils do not occur in the high-energy, cross-bedded sandstones of the Esplanade, although a few fusulinids and some bioclasts have been obtained in eastern Esplanade sandstones.

A large percentage of clay in the Manakacha Formation is believed to be of marine origin. Abundant illite associated with limestone occurs in the northwestern part of the area and some illite is found in scattered localities of eastern Grand Canyon. Corrensite, probably also of marine origin, has been found in various western areas, but is considered a product of low-energy deposition and is not associated with the abundant cross-stratified sandstones. Kaolinite is a common clay in eastern Grand Canyon, occurring among mudstones and siltstones, probably of continental origin.

In the Wescogame Formation, kaolinite is a common clay mineral among sandstones and siltstones throughout the eastern two-thirds of the area. Its abundance indicates that the majority of deposits in the formation are continental. Marine clays, including some illite, are largely confined to the extreme western part of the region. The interface of kaolinite occurrence with illite occurrence at various horizons in this formation possibly indicates shoreline locations at different stages of deposition.

[3] Determination of clay minerals and their genetic assignments for this study were by Phoebe Hauff, U.S. Geological Survey, Denver, Colorado.

The Esplanade Sandstone is characterized by an abundance of kaolinite and by a relative scarcity of illite. The kaolinite occurs principally in sandstone that composes a major part of the formation; the illite is confined to carbonate rocks, especially dolostone of the Pakoon Limestone. Illite is also associated with calcareous sandstone and evaporite deposits in the northwestern area. Its boundary with kaolinite farther east probably indicates a continental-marine interface. Corrensite is present only in the gypsum facies.

SANDSTONES — TEXTURE AND STRUCTURE

Some contrasts between the four formations of the Supai Group are clearly illustrated by the sandstone bodies of each. For example, the oldest formation, the Watahomigi, includes only a very small amount of sandstone, whereas the three upper formations contain large bodies of sandstone that compose major proportions of each formation. Sandstones in the upper formations, moreover, are cross-stratified on a medium to large scale and certain characteristics of the cross-strata serve to illustrate differences in the processes responsible for deposition.

Sandstone of the Watahomigi, in addition to being sparse, is composed almost exclusively of very fine sand grains, in contrast to the considerable amount of fine sand and scattered occurrences of coarser sand in the other formations. Furthermore, the proportion of sand deposits to silt deposits in the Watahomigi is about 1:3, whereas in the upper formations sand is 2 to 5 times as abundant as silt. In the Watahomigi, only small-scale cross-strata, consisting mostly of migrating ripple laminae and climbing ripples, but no large-scale cross-strata, are represented. Thus, textures and structures within the sand bodies indicate a low-energy environment for this formation.

In the Manakacha Formation, which, like the Watahomigi, also is considered marine, 72 percent of the samples show good sorting and 28 percent medium sorting. Mostly it consists of very fine grains, although many samples are dominantly fine grained and, in some localities, even coarser grains are not rare. The grain-size distribution pattern suggest bimodal sources. Cross-stratification is of medium to large scale throughout; some of it is extremely large scale in the central Grand Canyon area, with foresets ranging from 50 to 100 feet (15-30 m) in length, which is much greater than in the other formations. It consists entirely of tabular-planar and wedge-planar types; the trough type is not represented. The absence of troughs is one of its distinctive features.

The Wescogame Formation is dominated by a large sandstone body with mostly very fine rounded grains that are well sorted. However, it includes some coarse or very coarse particles in both eastern and western parts of the region, suggesting sources to the north and northwest, as also indicated by cross-strata dip-direction vectors. Unlike the other formations in the group, the Wescogame contains cross-strata not only of tabular- and wedge-planar types (fig. 100D,E,F), but also common festoons of large-scale trough types which suggest a fluvial origin. The ratio of sand to silt in the formation is 4:1.

In the Esplanade Sandstone, unlike the Wescogame and the Manakacha, the coarsest grains are largely in the central part of the area. The sand is mostly very fine grained and well sorted, and cross-strata dip-directions, with a spread of only 57°, trend southeast, which is more easterly than vectors of the underlying formations. The cross-strata are nearly half tabular-planar with the trough type (fig. 100B) uncommon. Foresets are both high and low angle and range from medium to large scale.

CARBONATE FACIES

Carbonate rocks in all formations of the Supai Group tend to be best developed and most extensive in the western part of the region and to be confined almost exclusively to the west in the upper two formations. Six principal facies are recognized in these formations as follows:

1. Aphanitic limestone or micrite.
2. Peloidal limestone, locally sandy, commonly cross-stratified.
3. Bioclastic or skeletal limestone, locally sandy, cross-stratified.
4. Accretal limestone that includes algal, oolitic, pisolitic structures.
5. Mixed-grain limestone, peloidal, bioclastic, accretal.
6. Dolostone, crystalline.

The distribution pattern of the facies in each formation contributes considerably to an understanding of the various depositional environments.

In the Watahomigi Formation, the carbonate unit in the middle is composed mostly of calcareous mudstone (micrite) in eastern and central areas, and of mixed-grained limestone in the west. The upper part of the formation in the far west is largely accretal limestone.

The middle part of the Manakacha in western and central Grand Canyon is a sandy, peloidal limestone that is highly cross-stratified, but this grades eastward into cross-stratified calcareous sandstone. Much of the upper part of the formation is cross-stratified peloidal sandstone, but in the far west, bioclastic limestone forms the uppermost part.

The major part of the Wescogame Formation is composed of medium- to large-scale, cross-stratified sandstone (fig. 100D,F), and carbonate rock is restricted to the western parts of the region. Limestone, which is sandy and cross-stratified (fig. 100E), contains much bioclastic material, but no micritic matrix; it grades into cross-stratified sandstone, so doubtless also represents high-energy conditions. Thin-bedded dolostone occurs only in the extreme northwestern part of the area.

The Esplanade Sandstone contains only negligible amounts of limestone throughout the Grand Canyon region, but near its western margin it intertongues into pure carbonate rock of the Pakoon Limestone. The Pakoon consists largely of dolostone, but is interbedded with bioclastic limestone (fig. 101B). This limestone is especially well represented in the western and southwestern parts of the area and characteristically contains a matrix of calcareous mud (micrite).

INTERPRETATION OF GENESIS

Although the Supai Group can properly be described as a series of red beds, the four formations of which it is composed represent at least four distinct depositional environments. Probably many more environments are included, for within each formation at least one transgression and regression is represented by alternating marine and continental deposits. This succession testifies to a very complex record of sedimentary change, both vertically and laterally.

The Watahomigi Formation, with its relatively large and widespread fauna of marine invertebrates and its extensive limestone deposits, clearly includes marine deposits throughout much of the Grand Canyon area. Brachiopods occur eastward as far as the type section of the formation at Havasu Canyon. Some calcareous mudstone (micrite) and a few bioclasts occur in easternmost localities. Further evidence of marine conditions is indicated by distribution of clay minerals —

marine types in western Grand Canyon, continental types in parts of the east. The environment must also have been one of low energy as indicated by the scarcity of sand, the lack of medium- or large-scale cross-strata, and the presence of small-scale structures such as ripple laminae and climbing ripples.

The Manakacha Formation, like the Watahomigi, is considered to be largely marine, but it represents high-energy conditions. Evidence of its marine environment is found 1) in the eastward extension of some limestone units from Nevada across the Grand Canyon area, 2) in the occurrence of bioclast accumulations and fusulinid tests scattered throughout the region, 3) in clay-mineral distribution with marine types common in western Grand Canyon and some illite also in scattered localities in the east, 4) by features of the large-scale cross-strata such as the exceptionally long foresets, the lack of trough-type structures, and the extensive gradation from cross-stratified sandstones into cross-stratified limestones. Best evidence of high-energy conditions is the very extensive medium- to large-scale cross-stratification.

The Wescogame Formation, although highly cross-stratified like the Manakacha and likewise reflecting high-energy conditions, differs from the Manakacha in certain major respects. The Wescogame contains relatively little carbonate rock except in the far west and is composed largely of cross-stratified sandstone. The cross-stratification does not have the extremely long foresets of the Manakacha, but does include a large amount of trough-type structure. Footprints of four-footed animals are preserved on foresets in numerous places. These features suggest fluviatile rather than marine origin, and the nature of the large-scale troughs, without associated climbing ripples above or lag material below, seems to indicate braided streams and not point-bar types. Clay minerals associated with the sandstones and siltstones consist mostly of kaolinite in the eastern and central areas, thus supporting the concept of a continental genesis.

The Esplanade Sandstone is difficult to interpret because available evidence is not conclusive. The medium-scale cross-strata that constitute a major part of the formation include some, but not an abundance of, trough structures. A large amount of kaolinite, but little illite, seems to indicate a nonmarine environment. On the other hand, gypsum beds in the northwestern area suggest lagoons, and the Pakoon Limestone that intertongues with the Esplanade from the west, with its fauna of corals, fusulinids, and other marine animals, shows that the sea lay in that direction. Divergent directions of cross-strata vectors and other bedding structures seem to indicate an estuarine depositional environment for most of the Esplanade Sandstone.

REFERENCES CITED

Darton, N. H., 1910, A reconnaissance of parts of northwestern New Mexico and northern Arizona: U.S. Geological Survey Bulletin 435, 88 p.

McKee, E. D., 1975, The Supai Group — Subdivision and nomenclature: U.S. Geological Survey Bulletin 1395–J, p. J1–J11.

McNair, A. H., 1951, Paleozoic stratigraphy of part of northwestern Arizona: American Association of Petroleum Geologists Bulletin, v. 35, p. 503–541.

PRELIMINARY REPORT OF BURIED VALLEYS AT THE MISSISSIPPIAN–PENNSYLVANIAN BOUNDARY IN WESTERN GRAND CANYON

George H. Billingsley, Jr.[1]

Abstract — Channel deposits of both marine and nonmarine sediments occur at the unconformity between Mississippian Redwall Limestone and Pennsylvanian Watahomigi Formation in western Grand Canyon. The valleys occupied by the channel deposits are cut into the Redwall Limestone and average 85 m in depth and 305 m in width. Marine fossils in the channel deposits are Chesterian age and record an additional marine transgression in latest Mississippian time prior to a major transgression of the sea at the beginning of Watahomigi deposition in Morrowan time.

INTRODUCTION

The unconformity between Mississippian and Pennsylvanian rocks in the Grand Canyon has been the subject of various studies during the last century. Throughout this time, the boundary between the two systems has been revised and refined and finally documented in the latest study by McKee and Gutschick (1969).

On the basis of faunal evidence the unconformity between the Mississippian Redwall Limestone and the overlying Pennsylvanian Watahomigi Formation represents a period of erosion or nondeposition through all or most of Late Mississippian and most of Early Pennsylvanian time (McKee and Gutschick, 1969). However, recent observations in the western Grand Canyon have revealed the presence of large-scale buried erosional valleys in the Redwall Limestone. These valleys average about 1,000 feet (305 m) in width and 280 feet (75 m) in depth. A maximum depth of 400 feet (122 m) occurs near Quartermaster Canyon (fig. 102). The valleys were filled with nonmarine and marine materials including fossils that establish a Late Mississippian (Chesterian) age and possibly very Early Pennsylvanian (Morrowan) age.

LOCATION AND DISTRIBUTION

The discovery of the buried valleys occurred during reconnaissance geological mapping of the western Grand Canyon in 1975 (Huntoon and Billingsley, 1977). The buried valleys are exposed in the walls of western Grand Canyon from Quartermaster Canyon eastward to Separation Canyon and Granite Park, a distance of approximately 36 miles (58 km). Eleven exposures have been examined to date and are listed with depth and width on the location map (fig. 103).

The major valley exposed in the Granite Park area can be traced westward to 209 Mile Canyon where it disappears under the Shivwits Plateau. The valley emerges from under the plateau near the junction of Green Spring and Surprise Canyons. Farther south at 217 Mile Canyon another major buried valley extends westward to 220 Mile Canyon and Separation Canyon. Both valleys merge near the vicinity of Surprise Canyon and continue on westward as one major valley to the Bat Tower overlook. The northern valley is designated the Granite Park Valley and the southern valley is designated the Separation Valley (fig. 104).

Figure 102. Exposed valley west side of Quartermaster Canyon. (Location 2 on fig. 103)

Smaller tributary valleys are found entering the Separation Valley near 220 Mile Canyon and west of Separation Canyon. In addition, two rather large tributaries enter the main valley west of Surprise Canyon, one from the north at Burnt Springs Canyon and one from the south at Quartermaster Canyon. It is not certain exactly where the main valley exits to the west. One large remnant valley is exposed near the mouth of the Grand Canyon at the Grand Wash Cliffs and may have been the exit area (fig. 104). Further study is needed to verify this. Recent canyon dissection has removed much of the buried valley system west of the Bat Tower overlook.

LITHOLOGY

Valley-fill sediments have been correlated from section to section on the basis of lithology, fossils, and color. They show lithologic continuity, but considerable variation in thickness. Drainage pattern, pebble imbrication, and distribution of the lower detrital sediments all indicate a westward-flowing drainage system.

The lower detrital deposits are considered fluvial and are composed largely of a basal chert conglomerate and calcareous sandstone. The conglomerate is composed of rounded white and yellow chert cobbles and pebbles 0.6 to 12 cm in diameter with a few boulders up to 70 cm in diameter. The conglomerate occurs in a matrix of poorly sorted coarse-grained

[1] Museum of Northern Arizona, Flagstaff, Arizona 86001.

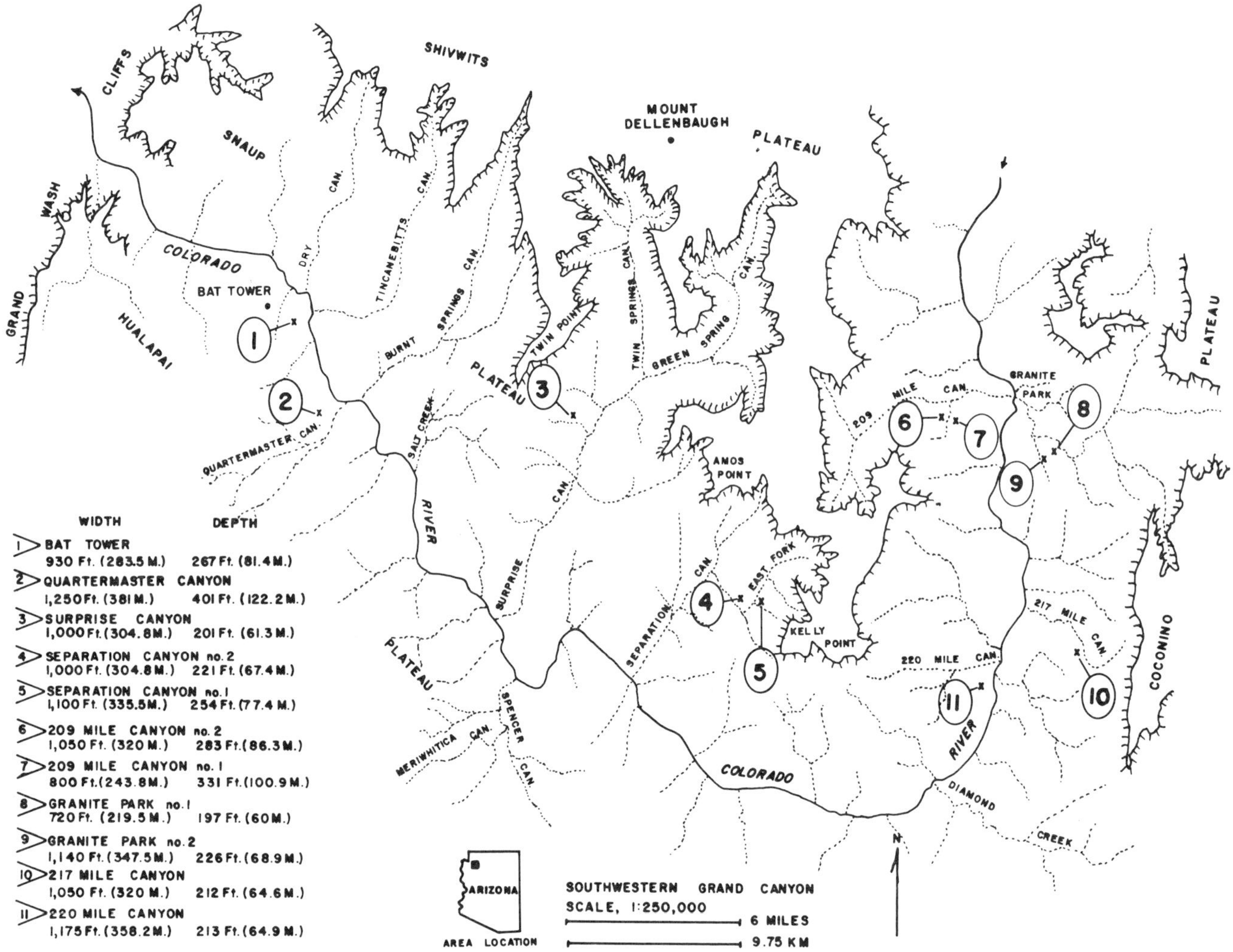

Figure 103. Location map of buried valley sections in the western Grand Canyon.

quartz sand. The cement includes siliceous, calcareous, and ferruginous types. A few of the chert pebbles and boulders contain fossils which are also found in the chert lenses and beds within the Mooney Falls Member of the Redwall Limestone. The conglomerate is associated with and overlain by a pale-yellow and dark-grey, calcareous sandstone that is medium- to coarse-grained and locally cross-stratified. Quartz grains are generally rounded. These deposits contain carbonaceous petrified wood and abundant carbonaceous plant remains. Thus, the lower detrital deposits are considered a continental deposit.

The upper part consists of three marine lithologic units. The lowest is a thin-bedded, purple-brown, gnarly coarse-grained limestone that forms a weak cliff. The middle unit is a slope-forming calcareous, red-brown and yellow sandstone. The upper unit is a cliff-forming relatively thin-bedded gray limestone. Both limestone units contain marine fossils; the sandstone between contains numerous burrows and cuspate ripple marks.

CORRELATION AND AGE

Brachiopods are the most abundant invertebrate fossils in the limestone units. In decreasing abundance, other fossils found are: mollusks, crinoid and blastoid remains, foraminifers, algae, trilobites, corals, and shark teeth. Plant and spore materials occur in the basal sandstone and conglomerates.

Brachiopod evidence of a Late Mississippian age for the valley-fill deposits is furnished by the "predominance of *Inflatia* and the presence of *Torynifer*" in the lower limestone unit (Mackenzie Gordon, Jr., 1978, written commun.). A specimen of *Glabrocingulum* is almost identical with a late Chesterian species from the Chainman Shale of Utah (Gordon, 1978, personal commun.). Foraminifers from the upper limestone represent uppermost Chesterian zone 19 of the Mamet scheme (Mamet and Skipp, 1970) according to Skipp (1978, written commun.). Ostracodes are present in all the limestone thin sections. Coral specimens found in the lower limestone were identified by W. J. Sando (1978, written commun.). One genus, *Barytichisma* sp., is relatively rare and has been described from the Morrowan of Texas, Arkansas, and Colorado and from the late Chesterian of Wyoming (Mamet zone 18 or 19). The genus is also found in beds of late Chesterian (zone 17) age in Idaho.

Further evidence of Late Mississippian age is supplied by spores extracted from the basal sandstone of the Granite Park Valley. R. M. Kosanke (1978, written commun.) reports *Convolutispora florida* and *Knoxisporites triradiatus* found in these samples. *Hadrohercas stereon* was also found and is known from the Goddard Formation of Oklahoma in the shale unit between the Rod Club and Overbrook Sandstone Members and the shale unit immediately above the Overbrook Sandstone Member. This interval is now considered to be Mississippian by Gordon (1967).

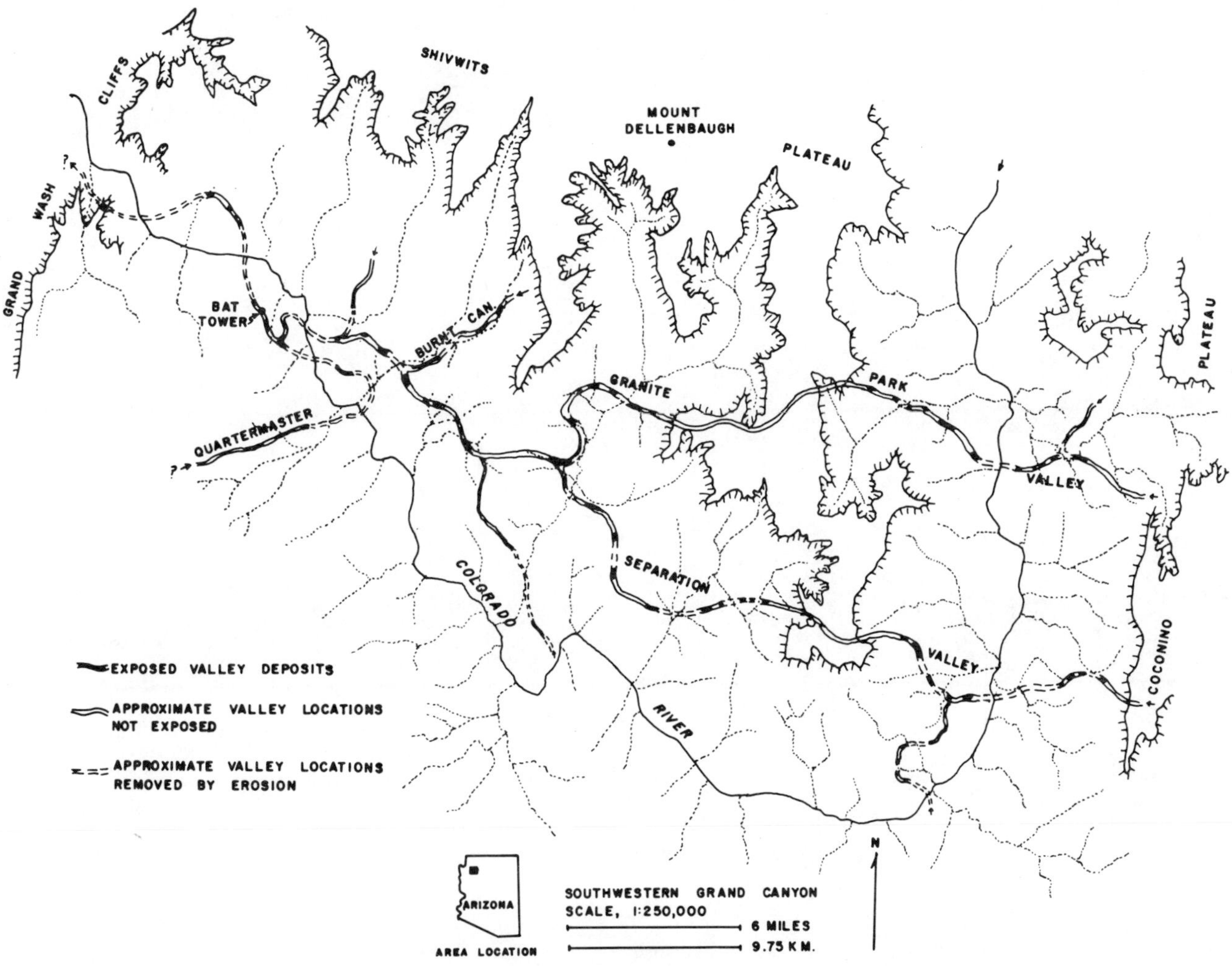

Figure 104. Location map of buried valley systems in the western Grand Canyon.

CONCLUSIONS

Evidence of regional uplift and warping of the Mississippian Redwall Limestone in western Grand Canyon was first noted by McKee and Gutschick (1969). They concluded that crustal uplift, probably warping, occurred at the time of initial marine transgression at the start of Supai deposition. This evidence is further supported by the presence of the buried valleys just described. Depths of the eastern buried valleys at Granite Park and 217 Mile Canyon show that regional uplift was between 200 (61 m) and 300 feet (91 m) above base level.

The valley-fill sediments containing flora and fauna of late Chesterian age indicate that warping or uplift of the western Grand Canyon region must have taken place in late Meramecian or early Chesterian time, followed by a gradual rising of sea level as the valleys were filled, first with continental and then marine sediments, prior to a major transgression of the sea during Morrowan time.

REFERENCES CITED

Gordon, Mackenzie, Jr., 1967, Contributions to stratigraphy: U.S. Geological Survey Bulletin 422–A, p. A54, A56–A57.

Huntoon, P. W., and Billingsley, G. H., 1977, 43 geologic maps of the western Grand Canyon: 7½ min. series, open file data, Grand Canyon Natural History Association, Grand Canyon, Arizona.

Mamet, B. L., and Skipp, B. A., 1970, Preliminary foraminiferal correlations of early Carboniferous strata in the North American Cordillera: Les Congrès et Colloques de l'Université de Liège, v. 55, p. 327-348.

McKee, E. D., and Gutschick, R. C., 1969, History of the Redwall Limestone of Northern Arizona: Geological Society of America Memoir 114, 726 p.

BIOSTRATIGRAPHY OF THE NACO FORMATION (PENNSYLVANIAN) IN CENTRAL ARIZONA

Douglas C. Brew[1]

Abstract — The Naco Formation (Pennsylvanian) in the Mogollon Rim region of central Arizona is a richly fossiliferous succession of dominantly marine limestones and interbedded fine clastics. The formation is thickest in east-central Arizona and thins northward, wedging out as it grades laterally and upward into the Supai Group (Pennsylvanian–Permian). The Naco is subdivided into three informal members: a thin basal member consisting chiefly of reddish insoluble residues that accumulated on an ancient karst topography developed on the underlying Redwall Limestone (Mississippian), a middle member consisting of alternating limestones and grayish shales or mudstones, and an upper member consisting of marine shales and limestones that intertongue with reddish clastics characteristic of the partly nonmarine Supai.

The Naco contains abundant and diverse faunas of foraminifers, coelenterates, bryozoans, brachiopods, mollusks, and echinoderms, plus other less common elements. Zonation is based chiefly on fusulinid foraminifers. The *Fusulina* Zone is represented by three subzones; only the latest Desmoinesian subzone is missing from the succession. The *Triticites* Zone includes forms indicative of both Missourian and Virgilian subzones. Thus, where it is thickest, the marine portion of the formation ranges in age from earliest Desmoinesian through latest Virgilian. Spiriferid, chonetid, and productid brachiopods, plus several other taxa are useful supplements to the fusulinid zonation.

The Naco Formation may be correlated with the Horquilla and Earp Formations of southeastern Arizona, the Hermosa Group of the Four Corners region, the upper Manakacha and the Wescogame Formations of the Grand Canyon, and the Callville Limestone of northwestern Arizona.

INTRODUCTION

The Naco Formation in central Arizona is a richly fossiliferous sequence of marine limestones and terrigenous clastic sedimentary rocks of Pennsylvanian age. The formation is well exposed in numerous outcrops within the Mogollon Rim region (fig. 105), a mountainous region underlain primarily by Precambrian and Paleozoic rocks that constitutes the "Transition Zone" between the Colorado Plateau and Basin and Range physiographic provinces (Wilson and Moore, 1959).

The Mogollon Rim, the prominent escarpment that constitutes the edge of the Colorado Plateau, is underlain by upper Paleozoic strata that dip gently northward and are relatively undeformed. Strata of the same age south of the rim, however, are cut by numerous faults, most of which appear to be high-angle normal faults whose displacements seldom exceed 125 m and most commonly measure less than 30 m. Although faulting thus introduces a greater degree of complexity to rocks south of the Mogollon Rim, the strata remain either horizontal or gently dipping, and readily lend themselves to stratigraphic examination.

The upper Paleozoic rocks in central Arizona were examined and reported on by Marvine (1875), Regan (1903), and Ransome (1916). The name Naco Limestone, first applied to Upper Carboniferous limestones in the vicinity of Bisbee, Arizona (Ransome, 1904), was extended to the Pennsylvanian strata of central Arizona by Huddle and Dobrovolny (1945). Several papers, notably those of Huddle and Dobrovolny (1952), Jackson (1951; 1952), McKee and Gutschick (1969), and Winters (1951; 1962; 1963), describe the stratigraphic relationships of the Naco Formation to the subjacent Redwall Limestone (Mississippian) and/or to the superjacent Supai Group. Brew (1965, Stratigraphy of the Naco Formation (Pennsylvanian) in central Arizona; Cornell University unpublished Ph.D. dissertation; 1970) reported on the stratigraphy and stratigraphic paleontology of the Naco in the Mogollon Rim region, and Brew and Beus (1976) published a study of a Naco fauna from the Kohl Ranch locality northeast of Payson. Ross (1972; 1973) published summaries of the Pennsylvanian and Early Permian depositional history of southeastern Arizona, including the Mogollon Rim region.

STRATIGRAPHY

In the Mogollon Rim region, the Naco Formation consists of predominantly marine sedimentary rocks of Pennsylvanian age. The formation disconformably overlies the Mississippian Redwall Limestone. This unconformity is a conspicuous regional datum and represents an ancient karst topography developed on the surface of the Redwall in Late Mississippian and Early Pennsylvanian time (Huddle and Dobrovolny, 1945; 1952; McKee and Gutschick, 1969; Brew, 1970). In situ terra rossa of the karst was partially reworked and incorporated into Pennsylvanian strata as the sea inundated the region from the southeast. The Naco Formation is approximately 365 m (1200 feet) thick in east-central Arizona where it ranges in age from earliest Desmoinesian through Virgilian. The formation thins northwestward and wedges out as it intertongues laterally and upward into the redbeds of the Pennsylvanian–Permian Supai Group, and measures only 125 m in thickness in the most northwesterly section measured, the Fossil Creek Canyon section (see fig. 105). Northwest of this locality the Naco Formation has not been recognized, although certain localities possess pockets of Naco-like lithologies preserved in topographically low areas of the ancient karst surface (D.P. Elston, 1978, personal commun.).

Brew (1970) subdivided the Naco Formation in the Mogollon Rim region into three members, informally termed (in ascending order) the alpha, beta, and gamma members.

[1] Department of Geosciences, University of Arizona, Tucson, Arizona 85721.

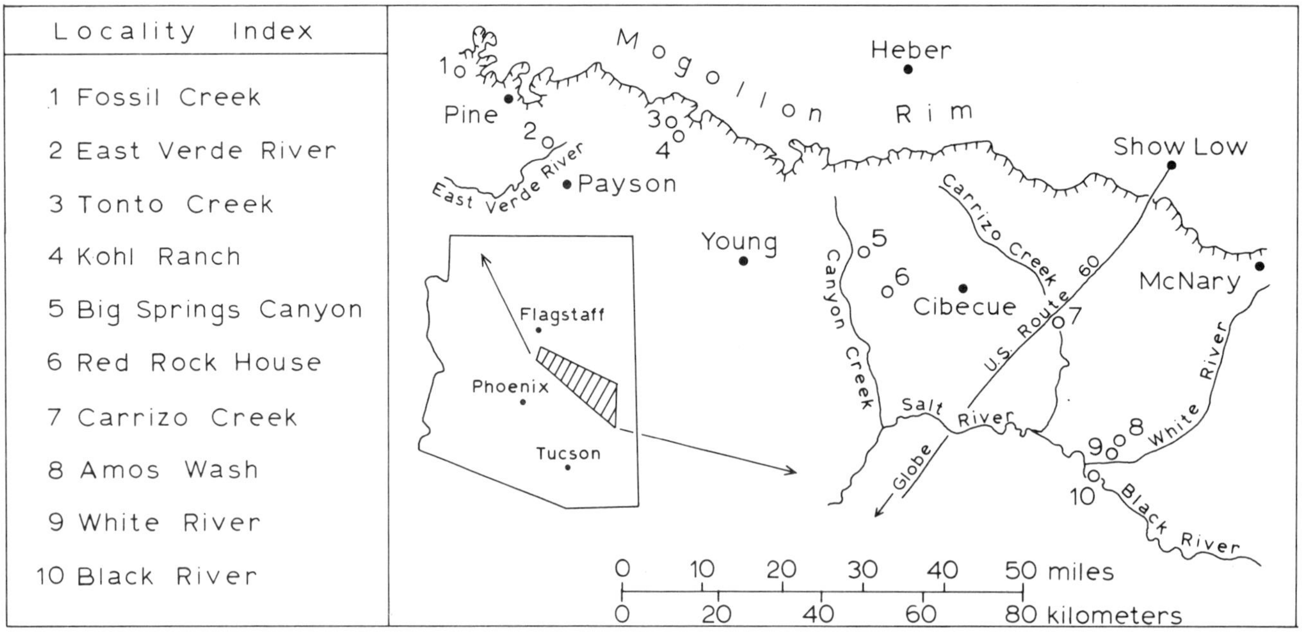

Figure 105. Locality map and index to sections.

Alpha member

The lower, or alpha, member ranges from 13 to 27 m (42 to 90 feet) in thickness and is composed chiefly of reddish-brown insoluble residues that accumulated on the karst topography developed on the Redwall. In most localities the member can be subdivided further into a basal nonbedded chert-rich conglomerate or breccia, siltstone, or mudstone, overlain by stratified mudstone, siltstone, or sandstone with local interbeds of fossiliferous limestone. The basal unit is interpreted to represent undisturbed residuum (Huddle and Dobrovolny, 1945; 1952; McKee and Gutschick, 1969; Brew, 1970). Moreover, since the basal in situ regolith contains silicified corals formerly indigenous to portions of the Redwall Limestone destroyed by solution, the uncommon opportunity is present to analyze the timing of events associated with development of the unconformity. Corals found in the residuum suggest that the youngest beds of the Redwall prior to solution were Late Mississippian, possibly Chesterian, in age; youngest Redwall strata untouched by solution are no younger than medial Mississippian, or Osagean, age (Sando, 1964). Fusulinid foraminifers found in fossiliferous interbeds of the upper portion of the alpha member suggest an earliest Desmoinesian age (*Fusulina hayensis* Subzone) in east-central Arizona, and an early Desmoinesian age (*Wedekindellina euthysepta* Subzone) in central Arizona (Brew, 1970). Thus it is clear that the shallow shelf sea in which Redwall carbonates accumulated persisted until the Late Mississippian or Chesterian time. Emergence of the shelf initiated development of karst, and during the hiatus that ensued, solution and erosion removed most of the upper Redwall units of Late Mississippian age. Invasion of the sea in early Desmoinesian time terminated the episode of karst formation.

Beta member

The middle, or beta, member is the most fossiliferous portion of the Naco Formation. At the time of maximum transgression, medial Desmoinesian, two reasonably well-defined facies can be recognized within the member: a more nearshore limestone and shale facies characterized by alternations of ledge-forming limestones and slope-forming grayish shales and mudstones, and a cliff-forming limestone sequence representing a depositional environment farther offshore. The member is approximately 256 m (840 feet) thick in east-central Arizona and thins northwestward, mimicking the regional thickness variation of the Naco as a whole. Both the basal and upper contacts with the alpha and gamma members, respectively, are transitional and characterized by interfingering successions of beds.

Deposition was apparently not continuous as sediments of the beta member accumulated, for numerous intraformational conglomerates, conglomerates, and local unconformities developed locally, especially at the close of the medial Pennsylvanian. The absence of fusulinid faunas characteristic of latest Desmoinesian and earliest Missourian age suggests that this was a time of widespread shoaling.

Gamma member

The upper, or gamma, member of the Naco ranges from 21 to 97 m in thickness (70-318 feet) and consists of a sequence of red or reddish-brown clastics interbedded with limestones that marks the transition between the largely nonmarine redbed facies of the Supai Group and the marine carbonate facies of the beta member. The unit is only sporadically fossiliferous, which, together with the reddish colors of the clastic units, serves to distinguish it from the limestone-clastic alternations of the beta member beneath. The gamma member as a whole is diachronous, apparently totally of Desmoinesian age in the Fossil Creek Canyon section and of Virgilian age in the Fort Apache Indian Reservation to the southeast; thus it clearly shows the time-transgressive southeastward progradation of the Supai clastics.

The Naco–Supai contact is therefore similarly time-transgressive, of late Desmoinesian age in the northwest and coincident with the Pennsylvanian–Permian boundary in east-central Arizona.

A diagrammatic summary of the stratigraphic relationships of the Pennsylvanian–Permian strata of central and east-central Arizona is shown as Figure 106. Major tectonic elements that influenced Pennsylvanian–Permian deposition in this region are shown on the index map included in this figure.

STRATIGRAPHIC PALEONTOLOGY

The Naco Formation contains abundant and diverse faunas that are typically dominated by brachiopods, but which also include foraminifers, corals, bryozoans, bivalves, gastropods, cephalopods, crinoids, echinoids, trilobites, and rare bone fragments. Zonation and correlation of the formation have been based chiefly on the stratigraphic distribution of fusulinid foraminifers; however, other taxa, most notably the spiriferid, chonetid, and productid brachiopods, hold the potential for further refinement of zonation.

Figure 107 presents a generalized stratigraphic column of the Naco Formation in east-central Arizona and the stratigraphic ranges of fusulinid and brachiopod species utilized in zonation. It should be noted that the stratigraphic ranges of some taxa, particularly the fusulinid foraminifers, seem unrealistically short, and it seems probable that additional collecting and analysis will alter this picture. The zones, subzones, and their diagnostic taxa are briefly summarized in the following sections.

Fusulina Zone

In east-central Arizona the lowest fusulinid-bearing beds contain *Fusulinella dosensis* Ross and Sabins, plus *Fusulina*,

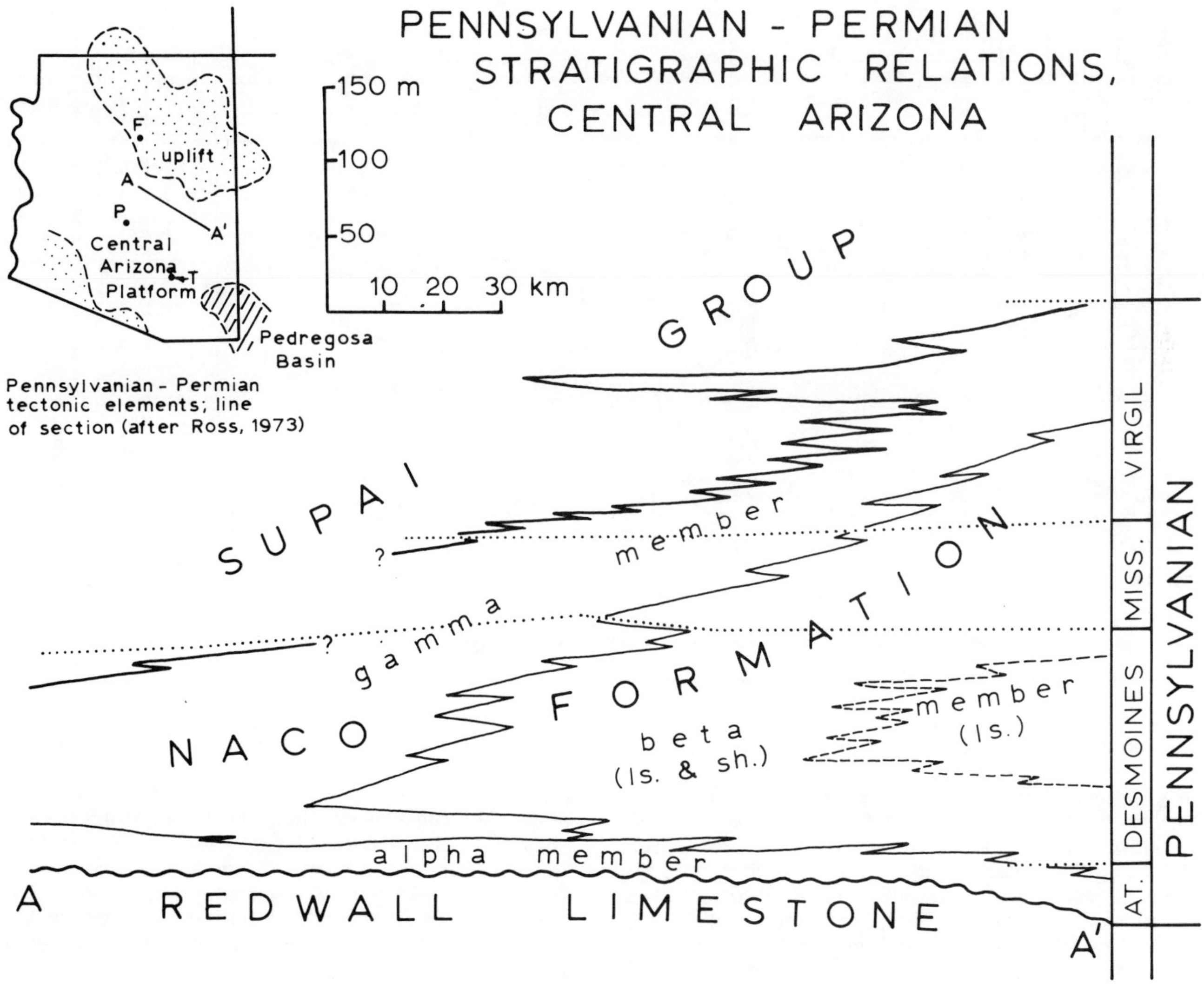

Figure 106. Diagrammatic sketch of Pennsylvanian – Permian stratigraphic relationships in the Mogollon Rim region; major tectonic elements (after Ross, 1973) of the Pennsylvanian – Permian of Arizona are shown on index map.

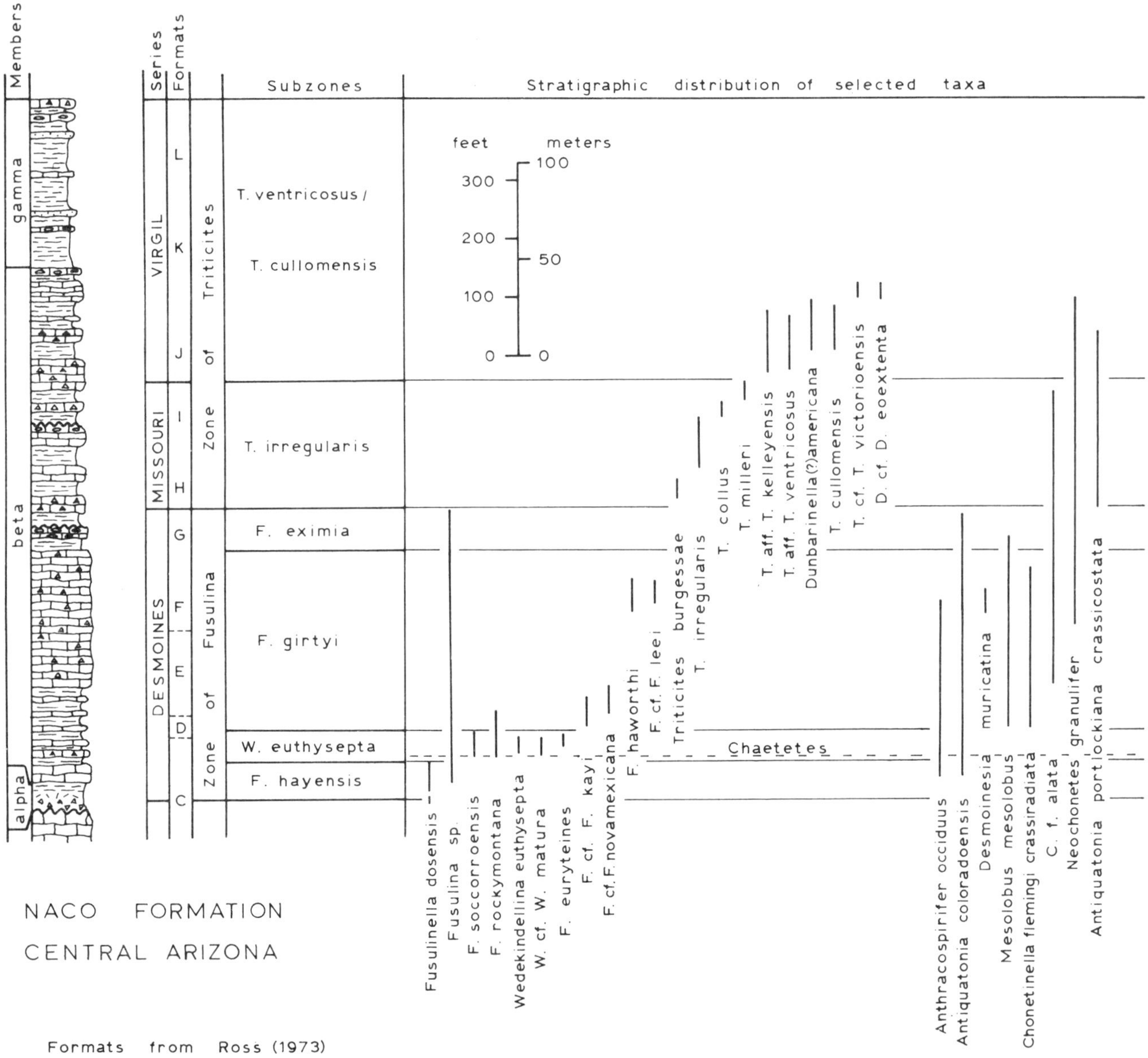

Figure 107. Generalized stratigraphic column and zonation of Naco Formation in east-central Arizona. *Chaetetes* horizon shown by dashed line.

an assemblage that suggests the lowest subzone of the Desmoinesian Series in Arizona, the subzone of *Fusulina hayensis* (= *Beedeina hayensis*; note that Ross in 1969 referred most North American species of *Fusulina* to the genus *Beedeina*; for convenience alone I have used the older designation of *Fusulina* throughout this paper). The *Wedekindellina euthysepta* Subzone is represented by *Fusulina socorroensis* Needham, *F. rockymontana* Roth and Skinner, *Wedekindellina euthysepta* (Henbest), and *W.* cf. *W. matura* Thompson. The remaining species of *Fusulina* included in Figure 107 characterize the *Fusulina girtyi* Subzone; however, two anomalous aspects to their distribution exist. First, the range of *Fusulina euryteines* Thompson, generally considered to be a more advanced form, appears to lie totally within the *W. euthysepta* Subzone. Second, specimens referred to *Fusulina leei* Skinner, which is supposed to be a lower Desmoinesian species, occur relatively high in the Desmoinesian section and are roughly contemporaneous with *Fusulina haworthi* (Beede), the most advanced form of the genus present in this region.

This distribution of species of *Fusulina* differs somewhat, where direct comparisons can be made, from that outlined by Ross (1973), who noted that *Fusulina leei* was found in the informal unit he termed Format D (see fig. 107 and following section on "Correlation"), *F. novamexicana* and *F. euryteines* in Format E, and *F. haworthi* in Format F. This discrepancy cannot be explained at the moment, but may be clarified by additional study.

No species of fusulinids diagnostic of the latest subzone of the Desmoinesian, or *F. eximia* Subzone, have been found in the succession. This fact, coupled with the sedimentologic evidence noted earlier, suggests that the late Desmoinesian was a time of widespread shoaling, during which either no deposition took place or fusulinid faunas could not penetrate the region.

Triticites Zone

The *Triticites irregularis* Subzone, which constitutes the lowermost subzone of the Upper Pennsylvanian and coincides with the Missourian Series, is marked by the presence of *Triticites burgessae* Burma, *T. irregularis* (Schellwein and Staff), *T. collus* Burma, and *T. milleri* Thompson. The overlying fusulinid faunas, characterized by *Triticites cullomensis* Dunbar and Condra, *T.* aff. *T. kelleyensis* Needham, *T.* aff. *T. ventricosus* (Meek and Hayden), and *Dunbarinella* (?) *americana* Thompson, are clearly Virgilian in age, yet cannot be separated into the *T. cullomensis* and *T. ventricosus* Subzones as is possible elsewhere. *Triticites* cf. *T. victorioensis* Dunbar and Skinner and *Dunbarinella* cf. *D. eoextenta* Thompson constitute the youngest fusulinid faunas within the region. These species are most commonly found in association with fusulinids characteristic of the *Pseudoschwagerina* Zone, the lowermost zone of the Permian. However, as the absence of forms diagnostic of this zone make a Wolfcampian age assignment untenable, the strata containing them are provisionally assigned a late Virgilian age.

Brachiopods

Several groups of brachiopods, especially the spiriferids, chonetids, and productids, have the potential to become important supplements to the fusulinid zonation. Although this is largely unrealized as yet, the potential is suggested by the stratigraphic ranges of the species shown in Figure 107. *Anthracospirifer occiduus* (Sadlick) is confined to the lower and middle Desmoinesian. Of three productid species, *Antiquatonia coloradoensis* (Girty) ranges throughout the Desmoinesian, *Desmoinesia muricatina* (Dunbar and Condra) occurs within the middle Desmoinesian, *Antiquatonia portlockiana* var. *crassicostata* Dunbar and Condra ranges from the Missourian into the Virgilian. The chonetid genera *Mesolobus, Chonetinella*, and *Neochonetes* display overlapping ranges, but in general are most characteristic of the Desmoinesian, Missourian, and Virgilian Series, respectively.

Other taxa

Of the remaining taxa, one may be singled out as being clearly of value in correlation within the Mogollon Rim region. *Chaetetes* sp. (?tabulate coral) is confined to a narrow stratigraphic interval within the lower Desmoinesian strata in the region as shown by the dashed line on Figure 107.

The remainder of the faunas are, in general, of lesser value for correlation between sections, either for reasons of their scarcity or their extended stratigraphic ranges. This does not, however, negate their value in inter-regional or inter-provincial correlation. Table 1 presents a faunal summary for the formation, exclusive of taxa already discussed or illustrated.

CORRELATION

Some general observations on correlation between the Pennsylvanian stratigraphic sections within the Mogollon Rim region and between the rim region and surrounding areas are necessary.

Ross (1973) recognized 15 unconformities in the Pennsylvanian–Early Permian succession in southeastern Arizona, 10 or 11 of which he also found within Pennsylvanian strata of the Mogollon Rim region. Two of the unconformities described by Ross, number 2, within the Derryan (Atokan), and number 8, at the Desmoinesian–Missourian boundary, were each considered to represent a hiatus of long duration (Ross, 1973). Ross interpreted all of the unconformities to be of regional extent, and used them to delineate a succession of informal stratigraphic units, or formats, suggesting that this subdivision into smaller informal stratigraphic units facilitated interpretation of the complex depositional and structural history. He noted that the scattered nature of fossil age determinations at that time and the uncertainty of their stratigraphic positions had not yielded a satisfactory time-stratigraphic framework for interpretation (Ross, 1973). While I agree with Ross that identification of thinner stratigraphic units is necessary to achieve a greater degree of resolution in correlation, I don't think that the unconformity-bounded units contribute significantly to this cause, particularly since Ross himself noted that lateral tracing and identification of formats and their bounding unconformities was made possible only by identification of fusulinids within the units (1973). Thus, emphasis on faunal analysis combined with careful stratigraphic control should yield positive results and produce a suitable time-stratigraphic framework for the region.

A second area worthy of comment concerns the nomenclature to be employed for the Pennsylvanian succession in central and east-central Arizona. Ross (1973) extended use of southeastern Arizona terminology into central Arizona, terming the Desmoinesian through roughly middle Virgilian portion of the Naco the Horquilla Formation and the uppermost portions the Earp Formation. Although a complete discussion of this change is inappropriate here, I prefer to retain the term Naco Formation for the central Arizona succession, for several reasons. First, extension of the names Horquilla and Earp does not materially improve the portrait of stratigraphic relationships in the Mogollon Rim region; second, the Horquilla and the Earp of southeastern Arizona differ somewhat lithologically from their proposed equivalents to the north; and third, the formations, as proposed, are of doubtful utility as map units. The Earp Formation, which would have to be interpreted as essentially equivalent to the gamma member, would pose especially difficult problems insofar as mapping is concerned.

In addition, Ross (1969) employed the name, Molas Formation, for what has herein been termed the alpha member. While recognizing that the alpha member clearly differs genetically from the remainder of the Naco, it is nevertheless difficult to set it apart as a separate formation because its thin and locally poorly exposed character renders it an unlikely candidate for a mappable unit. Moreover, while the similarity of the alpha member to the Molas Formation of the Four Corner's region is clearly established, continuity between these units is not. Thus, it seems more practical to treat this stratigraphic unit as a member of the Naco Formation.

The suggested correlation for the Naco Formation with strata of equivalent age in surrounding regions is shown in Figure 108.

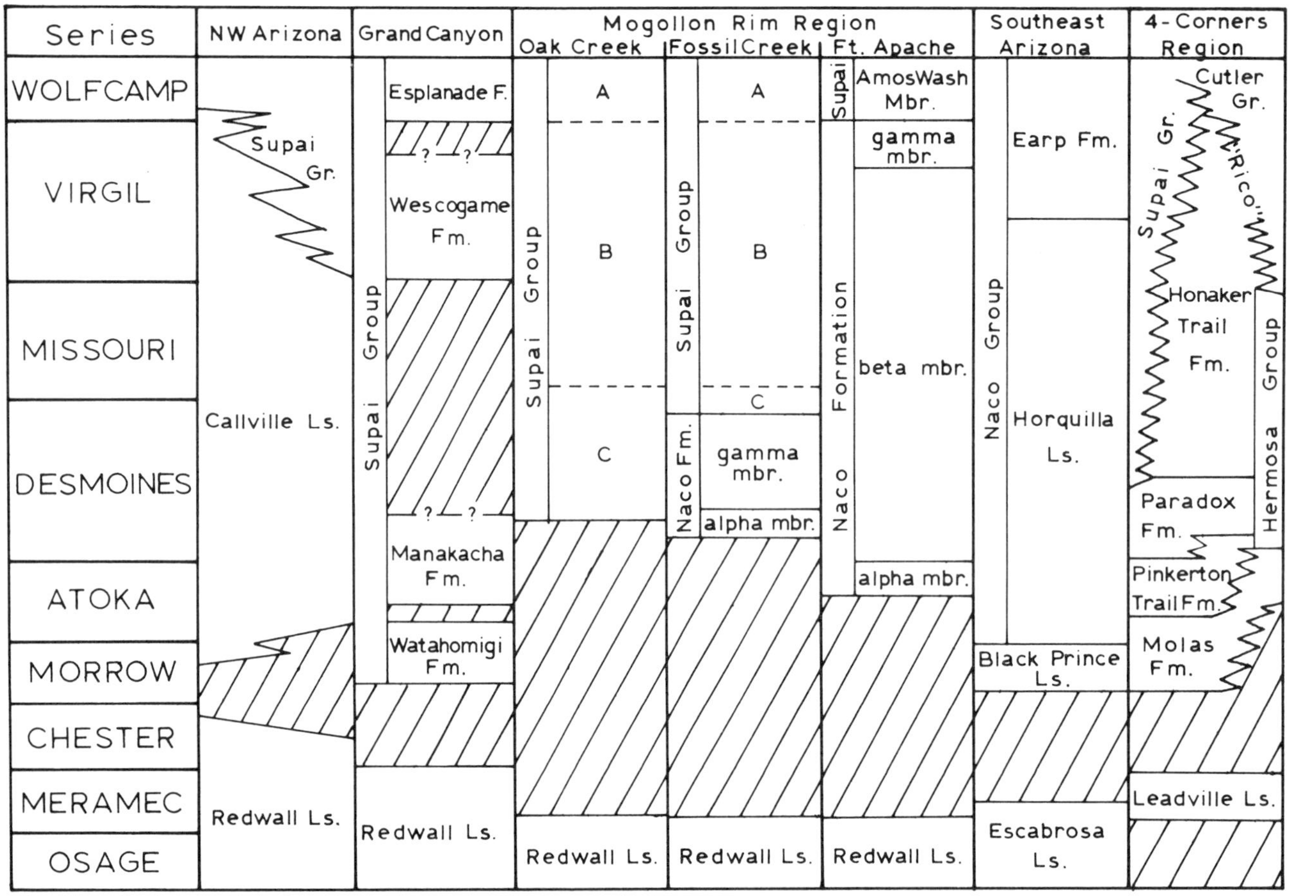

Figure 108. Correlation and stratigraphic nomenclature of Pennsylvanian units in Arizona and adjoining regions. Lettered subdivisions of Supai Group are older informal units used here pending extension of Grand Canyon nomenclature into Mogollon Rim region.

REFERENCES CITED

Brew, D.C., 1970, The Naco Formation (Pennsylvanian) in central Arizona: Plateau, v. 42, no. 4, p. 126–138.

Brew, D.C., and Beus, S.S., 1976, A Middle Pennsylvanian fauna from the Naco Formation near Kohl Ranch, central Arizona: Journal of Paleontology, v. 50, p. 888–906.

Huddle, J.W., and Dobrovolny, E., 1945, Late Paleozoic stratigraphy and oil and gas possibilities of central and northeastern Arizona: U.S. Geological Survey Oil and Gas Investigations Preliminary Chart 10.

———**1952,** Devonian and Mississippian rocks of central Arizona: U.S. Geological Survery Professional Paper 233–D, p. D67–D112.

Jackson, R.L., 1951, Stratigraphic relationships of the Supai Formation of central Arizona: Plateau, v. 24, no. 2, p. 84–91.

———**1952,** Pennsylvanian–Permian facies of the Supai Formation in central Arizona: Arizona Geological Society Guidebook for field excursions in southern Arizona, p. 143–146.

Marvine, A.R., 1875, Report on the geology of the route from St. George, Utah, to Gila River, Arizona: U.S. Geographical and Geological Surveys West 100th Meridian (Wheeler), v. 3, pt. 2, p. 189–225.

McKee, E.D., and Gutschick, R.C., 1969, History of the Redwall Limestone of northern Arizona: Geological Society of America Memoir 114, 726 p.

Ransome, F.L., 1904, The geology and ore deposits of the Bisbee Quadrangle, Arizona: U.S. Geological Survey Professional Paper 21, 168 p.

———**1916,** Some Paleozoic sections in Arizona and their correlation: U.S. Geological Survey Professional Paper 98–K, p. K133–K166.

Reagan, A.B., 1903, Geology of the Fort Apache region, Arizona: American Geologist, v. 32, p. 265–308.

Ross, C.A., 1969, Middle and Upper Pennsylvanian fusulinaceans, Gila Mountains, Arizona: Journal of Paleontology, v. 43, p. 1405–1422.

———**1972,** Pennsylvanian and Lower Permian episodes of epicontinental marine inundation, southeast Arizona: 24th International Geological Congress, Montreal, sec. 6, p. 343–348.

———**1973,** Pennsylvanian and Early Permian depositional history, southeastern Arizona: American Association of Petroleum Geologists Bulletin, v. 57, no. 5, p. 887–912.

Sando, W.J., 1964, Stratigraphic importance of corals in the Redwall Limestone, northern Arizona: U.S. Geological Survey Professional Paper 501–C, p. C39–C42.

Wilson, E.D., and Moore, R.T., 1959, Structure of Basin and Range province in Arizona: Arizona Geological Society Guidebook II, p. 89–105.

Winters, S.S., 1951, Permian stratigraphy in eastern Arizona: Plateau, v. 24, no. 1, p. 10–16.

———**1962,** Lithology and stratigraphy of the Supai Formation, Fort Apache Indian Reservation, Arizona: New Mexico Geological Society 13th Field Conference Guidebook, p. 87–88.

———**1963,** Supai Formation (Permian) of eastern Arizona: Geological Society of America Memoir 89, 99 p.

Table 1. — Faunal Summary of the Naco Formation, central Arizona

	Desmoinesian	Missourian	Virgilian
Coelenterata:			
Cladochonus	X	—	—
Multithecopora	—	—	X
Syringopora multattenuata	—	—	X
Canina	?	?	X
Lophophyllidium	X	—	X
Pseudozaphrentoides	X	—	X
Paraconularia kohli	X	—	—
Brachiopoda:			
Lingula	X	—	—
Petrocrania modesta	X	—	—
Cleiothyridina orbicularis	X	—	—
Crurithyris planoconvexa	X	—	X
Derbyia cf. *D. bennetti*	X	—	—
D. crassa	X	?	—
D. cf. *D. plattsmouthensis*	X	—	—
Beecheria ? cf. *B.* ? *bovidens*	X	—	—
Hustedia mormoni	—	—	X
Meekella cf. *M. striatocostata*	X	—	—
Condrathyris perplexa	X	—	X
Punctospirifer kentuckyensis	X	—	X
Rhipidomella carbonaria	X	—	—
Wellerella osagensis	—	—	X
W. osagensis immatura	X	—	—
Composita elongata	X	—	X
C. magna	—	—	X
C. ovata	X	X	X
C. subtilita	X	X	X
C. trilobata	?	X	X
(Brachiopoda cont. next col.)			
Brachiopoda (cont.):			
Neospirifer cameratus	?	—	—
N. dunbari	X	X	X
N. cf. *N. kansasensis*	X	X	X
N. latus	?	?	X
Antiquatonia hermosana	—	—	X
A. portlockiana	X	X	X
Echinaria semipunctata	X	—	X
Juresania nebrascensis	X	?	X
Kozlowskia splendens	X	X	?
Linoproductus cf. *L. meniscus*	X	—	—
L. platyumbonus	X	—	—
L. prattenianus	X	X	X
Retaria lasallensis	—	—	X
Bivalvia:			
Astartella	—	—	X
Aviculopecten	X	X	X
Aviculopinna	X	—	X
Myalina	X	X	X
M. (Myalina) nacoensis	X	—	—
M. (Orthomyalina)	—	?	X
Nucula cf. *N. wewokana*	—	—	X
Permophorus	X	X	—
P. subcostatus	X	—	—
Phestia	X	—	—
Pseudomonotis	—	—	X
Schizodus	X	X	?
Streblochondria	—	—	X
"Sedgewickia"	X	—	—
Wilkingia terminale	X	—	X
Gastropoda:			
Bellerophon (Bellerophon)	X	X	—
B. (B.) crassus	X	—	—
B. (Pharkidonotus) cf. *B. (P.) tricarinatus*	—	—	X
Euphemites	—	X	X
Glabrocingulum ?	—	—	X
Goniasma	—	—	X
Ianthinopsis	—	X	—
Knightites (Cymatospira)	—	—	X
K. (Retispira)	—	X	X
Naticopsis	X	—	?
Paragoniozona	—	—	X
Platyceras (Platyceras) parvum	X	—	—
Pseudozygopleura	—	X	X
Straparollus (Amphiscapha) subrugosus	X	—	X
S. (Euomphalus)	X	—	X
Cephalopoda:			
Mooreoceras	—	—	X
Tainoceras cf. *T. monilifer*	—	—	X
Titanoceras	X	—	—
Crinoidea:			
Delocrinus	X	—	—
Eirmocrinus jeani	X	—	—
Endelocrinus	X	—	—
Paradelocrinus nederi	X	—	—
Parethelocrinus	X	—	—
Sciadiocrinus brewi	X	—	—
Synarmocrinus carrizoensis	X	—	—

BIOSTRATIGRAPHY OF THE UPPER CARBONIFEROUS (PENNSYLVANIAN) IN THE VIRGIN MOUNTAINS OF SOUTHERN NEVADA AND NORTHWESTERN ARIZONA

Walter H. Pierce[1]

Abstract — Strata of the Virgin Mountains area, within the transition between the Supai Group and the Bird Spring Formation, contain brachiopods, conodonts, corals, and fusulinids indicative of Chesterian, Morrowan, Atokan, Des Moinesian, Virgilian, and Wolfcampian time.

INTRODUCTION

Mississippian, Pennsylvanian, and Permian stratigraphy of the Virgin Mountains (fig. 109) is critical to the correlation of the Supai Group of the Grand Canyon with the Bird Spring Formation of the Cordilleran thrust belt of southern Nevada (fig. 110). The Virgin Mountains lie between the Las Vegas Line of Welsh (1959) and the physiographic boundary that separates the Colorado Plateau from the Basin and Range Province (fig. 110). Longwell (1921, 1928), McNair (1951), Longwell et al. (1965), Welsh (1959), Bissell (1962, 1969), and Brill (1963) have also presented biostratigraphic data which pertain to the problem. The data presented herein augment their data.

Acknowledgements

George J. Verville of AMOCO kindly identified and interpreted the age of the fusulinids in the report. Edwin D. McKee and Richard H. DeVoto provided helpful comments on the original research.

BIOSTRATIGRAPHIC COLLECTIONS

The following collections aid in the correlation and have enabled the author to assign ages to lithostratigraphic units. The collections come from three stratigraphic sections, (fig. 109):

1. Grand Gulch section, Arizona, Mohave County, secs. 9, 10, and 11, T. 34 N., R. 14 W.
2. Pakoon Ridge section, Arizona, Mohave County, secs. 15 and 16, T.18 S., R. 69 E.
3. Lime Ridge section, Nevada, Clark County, secs. 2, 10, and 11, T. 18 S., R. 69 E.

The fossil collection list includes a code letter which shows the stratigraphic location on Figure 111, a stratigraphic distance above the Redwall Limestone, a name of the lithostratigraphic unit from which the collection was made, a faunal list, and an interpretation of the age.

A. Lime Ridge section, about 3 feet (1 m) below top of Redwall Limestone. Redwall Limestone.

Conodonts:
Spathognathodus sp.
Synprioniodina denticamura
Hindeodus sp.

Interpretation: Meramecian.

B. Lime Ridge section, 6 feet (1.83 m) above top of Redwall Limestone, within Battleship Wash Formation.

Conodonts:
Ozarkodina curvata
Magnilaterella robusta
Spathognathodus scitulus
Spathognathodus rexroadi
Synprioniodina denticamura
Hibbardella sp.
Lambdagnathus fragilidens

Interpretation: Chesterian, equivalent of Beech Creek to Glen Dean of type Chesterian, or the *Gnathodus bilineatus–Cavusgnathus altus* and/or *Gnathodus bilineatus–Kladognathus mehli* Zone of Collinson et al (1971).

C. Lime Ridge section, 194–201 feet (59.0–71.3 m) above top of Redwall Limestone, within lower member of Callville Limestone.

Conodonts:
Adetognathus lautus
Neognathodoides bassleri symmetricus
Idiognathodus sinuosis
Streptognathodus sp.
Hindeodella sp.

Interpretation: lower middle Morrowan.

D. Lime Ridge section, 233–236 feet (71.9–71.9 m) above top of Redwall Limestone, within lower member of Callville Limestone.

Conodonts:
Adetognathus lautus
Neognathodoides bassleri symmetricus
Idiognathodus sinuosis
Hindeodella sp.

Interpretation: middle Morrowan.

E. Lime Ridge section, 333–336 feet (117–118 m) above the top of the Redwall Limestone, within lower member of Callville Limestone.

Corals:

---

[1] Department of Geology, Ball State University, Muncie, Indiana 47306.

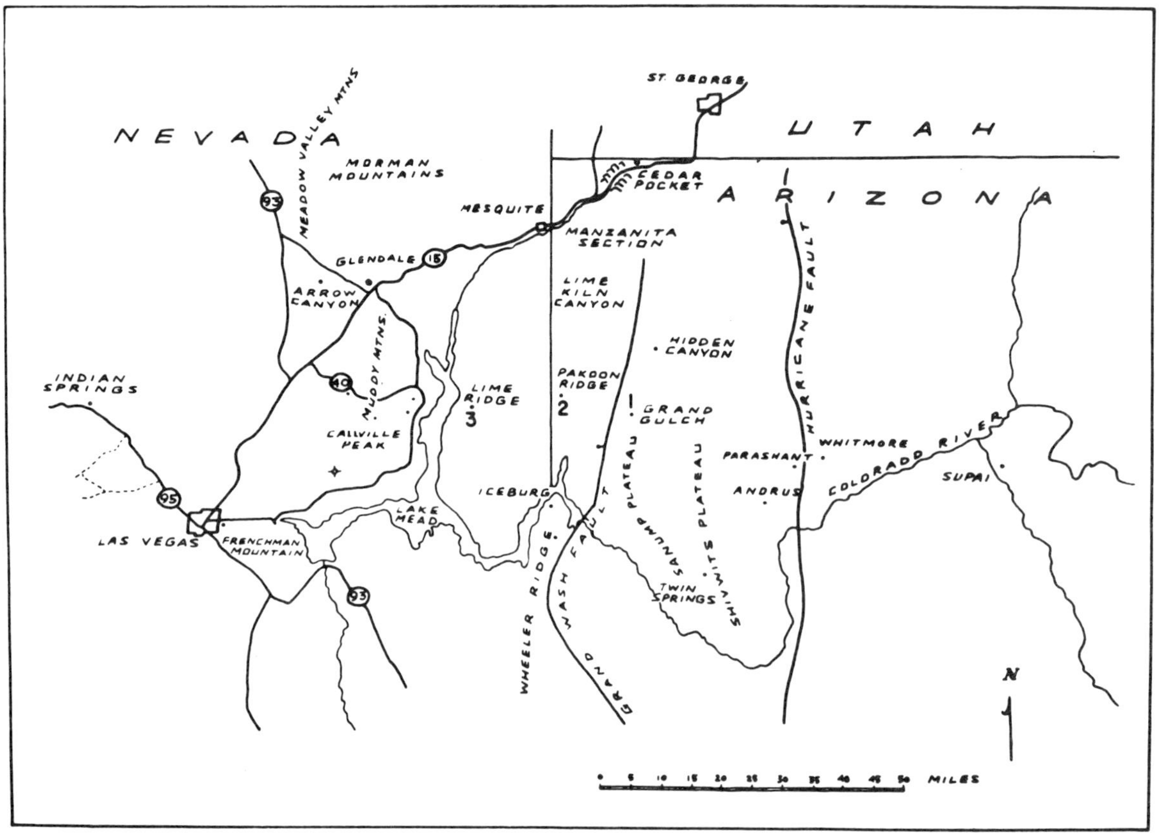

Figure 109. Index map of Virgin Mountain region showing location of stratigraphic sections.

Chaetetes sp.

Interpretation: Atokan.

F. Lime Ridge section, 418–422 feet (127–129 m) above top of Redwall Limestone, within lower member of Callville Limestone.

Corals:

Chaetetes sp.

Interpretation: Atokan

G. Lime Ridge section, 574–576 feet (174–176 m) above top of Redwall Limestone, within unnamed sandstone member of Callville Limestone.

Fusulinids:

Fusulina sp.

Interpretation: Des Moinesian.

H. Lime Ridge section, 584–590 feet (178–180 m) above top of Redwall Limestone, within upper member of Callville Limestone.

Brachiopods:

Anthracospirifer curvilateralis chavezae

Interpretation: "*A. curvilateralis chavezae* is one of the most common brachiopods in the lower and middle parts of the Des Moinesian interval in the southern Sangre de Cristo area. This species also occurs in the upper part of the Atokan sequence." (Sutherland and Harlow, 1973, p. 83).

I. Lime Ridge section. This sample was placed in the measured stratigraphic section by physical correlation across a major fault from sec. 2, T.18 S., R.69 E. It occurs 6 feet (2 m) below a conglomerate which approximately correlates with Esplanade conglomerate (McKee, 1975) and therefore 748 feet (228 m) above top of Redwall Limestone within upper member of the Callville Limestone.

Fusulinids:

Triticites sp.

Interpretation: George J. Verville identified these specimens and made the following comments — "These specimens could be either upper Virgilian or lower Wolfcampian. I favor a lower Wolfcampian assignment."

J. Pakoon Ridge section, 15–16 feet (4.57–4.88 m) above top of Redwall Limestone, within lower member of Watahomigi Formation.

Brachiopod:

Craniops implicata

Interpretation: Chesterian or Morrowan.

K. Pakoon Ridge section, 34–35 feet (10.4–10.7 m) above top of Redwall Limestone, within lower member of Watahomigi Formation.

Conodonts:

Rhachistognathus primus
Rhachistognathus muricatus
Adetognathus lautus
Adetognathus spathus
Adetognathus unicornis
Ozarkodina curvata
Neognathodus bassleri symmetricus
Spathognathus sp.

Hindeodella sp.

Interpretation: *Rhachistognathus primus* Zone of Lane and Straka (1974) of lowermost Morrowan.

L. Pakoon Ridge section, 132 feet (40.2 m) above top of Redwall Limestone, within lower member of Callville Limestone.

Conodonts:
Adetognathus lautus
Adetognathus spathus
Idiognathus sinuosis
Neognathodus medexultimus
Hibbardella sp.
Geniculatus sp.
Lonchodina sp.
Hindeodella sp.
Neognathodus bassleri symmetricus

Interpretation: *Idiognathodus sinuosis* Zone of Lane and Straka (1974) or upper middle Morrowan.

M. Pakoon Ridge section, 189–190 feet (57.6–58.2 m) above top of Redwall Limestone, within lower member of Callville Limestone.

Brachiopods:
Anthracospirifer sp.
Linoproductus cf. *L. platyumbonus* Sutherland and Harlow (1973)
Linoproductus nodosus
Linoproductus sp.
Pulcratia picuris Sutherland and Harlow (1973)

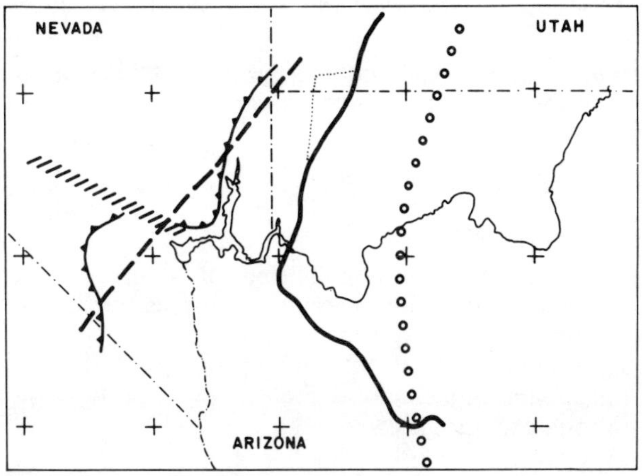

Figure 110. Map showing regional tectonic "lines" and physiographic boundaries mentioned in paper. Circles – Wasatch Line of Kay (1951); solid line – physiographic boundary between Colorado Plateau and Basin and Range Province (Fenneman and Johnson, 1946) with modification in dotted line by Hunt (1956); thrust belt symbol marks front of Cordilleran Thrusting; hashured line – Walker Lane or Las Vegas Shear zone.

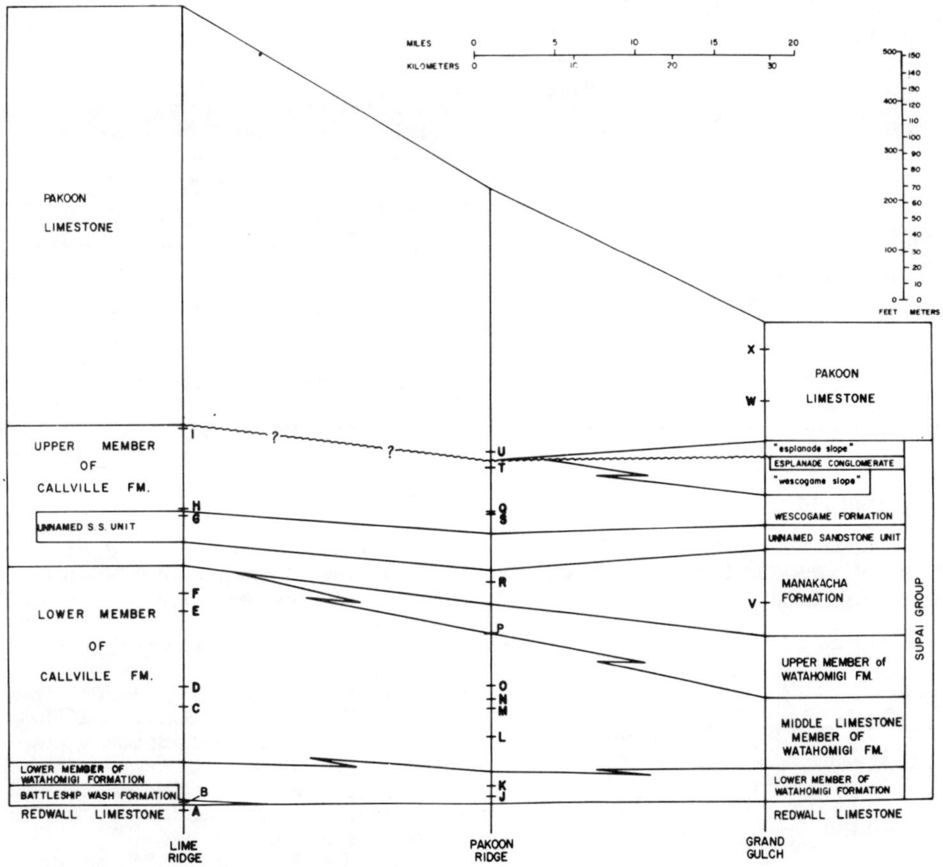

Figure 111. Stratigraphic diagram showing relation of biostratigraphic collections (indicated by letters) to formations and members.

Interpretation: "*Linoproductus nodosus* is widely distributed in rocks of Morrowan age." (Sutherland and Harlow, 1973).

N. Pakoon Ridge section, 206–208 feet (62.8–63.4 m) above top of Redwall Limestone, within lower member of Callville Limestone.

Conodonts:
Adetognathus lautus

Interpretation: Morrowan or Atokan

O. Pakoon Ridge section, 332–335 feet (101–102 m) above top of Redwall Limestone, within lower member of Callville Limestone.

Conodonts:
Idiognathus sinuosis
Neognathodus bassleri bassleri
Neognathodus sp.
Adetognathus lautus
Hibbardella sp.
Ozarkodina sp.
Hindeodella sp.

Interpretation: Morrowan or Atokan

P. Pakoon Ridge section, 335–338 feet (102–103 m) above top of Redwall Limestone, at top of lower member of Callville Limestone.

Corals:
Chaetetes sp.

Interpretation: Atokan.

Q. Pakoon Ridge section, data from Brill (1963, fig. 9) measured from diagram approximately 133 feet (40.5 m) above *Chaetetes* sp. or 471 feet (144 m) above top of Redwall Limestone. By inspection of lithologies collection appears to be 580 feet above Redwall Limestone, within upper member of Callville Limestone.

Fusulinids:
Triticites beedei
Triticites cf. *T. plummeri*

Interpretation: Virgilian.

R. Pakoon Ridge section, data from Welsh (1959) approximately 440–450 feet (134–137 m) above top of Redwall Limestone, appears to be within lower member of Callville Limestone.

Fusulinids:
Eoshubertella sp.
Fusulina sp.
Wedekindellina sp.

Interpretation: Des Moinesian.

S. Pakoon Ridge section, data from Welsh (1959) approximately 578 feet (176 m) above top of Redwall Limestone, within upper member of Callville Limestone.

Fusulinids:
Triticites sp.

Interpretation: Virgilian.

T. Pakoon Ridge section, 666–671 feet (203–205 m) above top of Redwall Limestone, within upper member of Callville Limestone.

Fusulinids:
Triticites sp.

Interpretation: Specimens identified by George J. Verville and interpreted to be Virgilian in age.

U. Pakoon Ridge section, 699 to 700 feet (213 m) above top of Redwall Limestone, within Pakoon Limestone 15 feet above Esplanade conglomerate.

Fusulinids:
Triticites sp.

Interpretation: George J. Verville identified these specimens and made the following comment: "These specimens could be either Virgilian or lower Wolfcampian, but I would rather assign them to the upper Virgilian."

V. Grand Gulch section, 394–396 feet (120–121 m) above top of Redwall Limestone, within Manakacha Formation.

Fusulinids:
Fusulinella sp.

Interpretation: Identified and determined to be Atokan by George J. Verville.

W. Grand Gulch section, 799–804 feet (244–245 m) above top of Redwall Limestone, within Pakoon Limestone.

Fusulinids:
Schwagerina sp.

Interpretation: Identified and determined to be Wolfcampian by George J. Verville.

X. Grand Gulch section, 903–904 feet (275–276 m) above top of Redwall Limestone, within Pakoon Limestone 50 feet (15 m) below Sanup Plateau.

Fusulinids:
Schwagerina linearis

Interpretation: Identified and interpreted by George J. Verville to be upper Wolfcampian.

CONCLUSIONS

Figure 111 shows the author's lithostratigraphic correlation in relation to the biostratigraphic collections. The correlation attempts to show the relationship of the Supai Group of the Grand Canyon (McKee, 1975) and the Callville Limestone and Pakoon Limestone of the Virgin Mountain Area as used by McNair (1951) and Lumsden et al. (1973).

REFERENCES CITED

Bissell, H.J., 1962, Pennsylvanian and Permian rocks of Cordilleran area. *In* Branson, C.S. (ed.), Pennsylvanian System in the United States: American Association of Petroleum Geologists Symposium, Tulsa, Oklahoma, p. 188–263.

———**1969,** Permian and Lower Triassic transition from shelf to basin (Grand Canyon Region): Four Corners Geological Society Guidebook, p. 135–169.

Brill, K.G., 1963, Permo-Pennsylvanian stratigraphy of western Colorado Plateau and eastern Great Basin regions: Geological Society of America Bulletin, v. 74, p. 307–330.

Collinson, C.W., Rexroad, C.B., and Thompson, Thomas L., 1971, Conodont zonation of the North American Mississippian. *In* Conodont Biostratigraphy: Geological Society of America Memoir 127, p. 353–394.

Fenneman, N.M., and Johnson, D.W., (in coop. with Physiographic Comm., U.S. Geological Survey) **1946,** Physical divisions of the United States: U.S. Geological Survey, scale 1:7,000,000.

Hunt, C.B., 1956, Cenozoic geology of the Colorado Plateau: U.S. Geological Survey Professional Paper 279, 99 p.

Kay, M. 1951, North American Geosynclines: Geological Society of America Memoir 48, 148 p.

Lane, H.R., and Straka, J.J., 1974, Late Mississippian and Early Pennsylvanian conodonts, Arkansas and Oklahoma: Geological Society of America Special Paper 152, 144 p.

Longwell, C.R., 1921, Geology of the Muddy Mountains, Nevada, with a section to the Grand Wash Cliffs, in western Arizona: American Journal of Science, v. 201, p. 39–62.

——**1928,** Geology of the Muddy Mountains, Nevada, with a section through the Virgin Range to the Grand Wash Cliffs, Arizona: U.S. Geological Survey Bulletin 798, 152 p.

——**Pameyan, E.H., Bowyer, B., and Roberts, R.J., 1965,** Geology and mineral deposits of Clark County, Nevada: Nevada Bureau of Mines Bulletin 62, p. 218.

Lumsden, D.N. et al., 1973, Lithostratigraphic analysis of the Bird Spring–Callville Group and Pakoon Formation (Pennsylvanian–Lower Permian), southern Clark County, Nevada: Journal of Sedimentary Petrology, v. 43, p. 655–671.

McKee, E.D., 1975, The Supai Group — Subdivision and Nomenclature: U.S. Geological Survey Bulletin 1395–J, 11 p.

McNair, A.H., 1951, Paleozoic stratigraphy of part of northwestern Arizona: American Association of Petroleum Geologists Bulletin, v. 35, p. 503–541.

Sutherland, P.K., and Harlow, F.H., 1973, Pennsylvanian Brachiopods and Biostratigraphy in Southern Sangre De Cristo Mountains, New Mexico: New Mexico Bureau of Mines and Mineral Resources Memoir 27, 173 p.

Welsh, H.E., 1959, Biostratigraphy of the Pennsylvanian and Permian System of southern Nevada: University of Utah unpublished Ph.D. dissertation, Salt Lake City, Utah, 106 p.

LEXICON OF PALEOZOIC STRATIGRAPHIC NAMES USED IN NORTHWESTERN ARIZONA AND SOUTHEASTERN NEVADA

Compiled by T. L. Klockenbrink[1]

This compilation was derived from available lexicons, catalogues, and other available published literature. Below each rock unit and its age are listed:

1. Areal distribution.
2. First reference in literature.
3. Type locality.
4. Generalized lithology and thickness.
5. General information and depositional environment.

Anchor Limestone (of Monte Cristo Group) — L. Mississippian

1. SE Nevada.
2. Hewett, 1931, U.S.G.S. Prof. Paper 162, p. 10.
3. Anchor Mine, Goodsprings quad., Nevada.
4. Massive gray ls. with thin chert layers, some dol. (65–400′).
5. Overlies Dawn Limestone; underlies Bullion Dolomite (both of Monte Cristo Group).

Arrow Canyon Formation — Devonian

1. Clark County, Nevada.
2. Langenheim et al., 1962, AAPG Bull., v. 46, p. 592–609.
3. Arrow Canyon Range, Clark County, Nevada.
4. Ls. with interbedded ss. (1100′).
5. Overlies Moapa Fm.; underlies Crystal Pass Ls.

Arrowhead Limestone (of Monte Cristo Group) — M. Mississippian

1. SE Nevada.
2. Hewett, 1931, U.S.G.S. Prof. Paper 162, p. 9.
3. Arrowhead prospect, Goodsprings quad., Nevada.
4. Alternating layers of thin-bedded blue and gray ls. and gray sh.; highly fossiliferous (10–20′).
5. Overlies Bullion Dolomite; underlies Yellowpine Limestone (both of Monte Cristo Group).

Banded Mountain Member (of Bonanza King Fm.) — M. to U. Cambrian

1. SE Nevada.
2. Barnes and Palmer, 1961, U.S.G.S. Prof. Paper 424–C, p. C–101.
3. Banded Mt., Nye County, Nevada.
4. Sequence of ls. and dol. (2500′).
5. Upper member of Bonanza King Fm.; overlies Papoose Lake Member.

Battleship Wash Formation — Meramec - Chester

1. SE Nevada.
2. Langenheim and Langenheim, 1965, Illinois Acad. Sci. Trans., v. 58, p. 225–240.
3. Arrow Canyon Range, Clark County, Nevada.
4. Ss., sandy ls., and ls. (80′).
5. Placed in Bird Spring Group by Langenheim et al. (1965); Webster et al. (1967) stated it was not related to Bird Spring Group or Indian Springs Fm.; underlies Indian Springs Fm.

Bird Spring Formation — U. Mississippian - L. Permian

1. Nevada, California, and Utah.
2. Hewett, 1931, U.S.G.S. Prof. Paper 162, p. 9–21.
3. Bird Spring Range, Nevada.
4. Limestone (1000–7000′).
5. Overlies Monte Cristo Group (marine).

Black Prince Limestone — U. Mississippian or L. Pennsylvanian

1. SE Arizona.
2. Romslo, 1949, U.S. Bur. Mines Rept. Inv. 4505, p. 5.
3. Black Prince Mine, Cochise County, Arizona.
4. Basal red-purple sh. with chert and ls. nodules and upper pink ls. (120–170′).
5. Underlies Horquilla Ls. and overlies Escabrosa Ls.

Bonanza King Formation — M. to U. Cambrian

1. SE California and Clark County, Nevada.
2. Hazzard and Mason, 1936, G.S.A. Bull., v. 47, p. 229–240.
3. Bonanza King Mine, Providence Mts., California.
4. Sandy dol., sh., light-gray dol. (2000′±).
5. Contains two members, in ascending order: Papoose Lake Mbr. and Banded Mountain Mbr.

Bright Angel Shale (of Tonto Group) — M. and U. Cambrian

1. Northern Arizona, SE Utah, SW Colorado, NW New Mexico.
2. Noble, 1914, U.S.G.S. Bull. 549.
3. Bright Angel Canyon, Coconino Co., Arizona.
4. Soft greenish micaceous and foss. sh. and ss., with brn. crystalline ls. (25–375′).
5. Overlain conformably by Muav Ls.; underlain conformably by Tapeats (Ignacio) Ss. (marine).

Bullion Dolomite (of Monte Cristo Group) — L. Mississippian

1. SE Nevada.
2. Hewett, 1931, U.S.G.S. Prof. Paper 162, p. 10.
3. Bullion Mine, Nevada.
4. Massive gray ls., altered to cream-colored and white dol. (185–300′).
5. Underlies Arrowhead Limestone; overlies Anchor Limestone (both of Monte Cristo Group).

Callville Limestone — Pennsylvanian

1. SE Nevada and NW Arizona.
2. Longwell, 1921, Am. Jour. Sci., 5th, v. 1, p. 47.
3. Callville Mt., Clark County, Nevada.
4. Gray ls. and interbedded slts. in upper part (300–2000′).
5. Overlies Redwall Fm.; intertongues with lower part of Supai Group to east (marine).

---

[1] Department of Geology, Northern Arizona University, Flagstaff, Arizona 86011.

Chisholm Shale — M. Cambrian

1. Eastern Nevada.
2. Walcott, 1916, Smithsonian Misc. Coll., v. 64, No. 5, Pub. 2420, p. 409–410.
3. Chisholm Mine, NW of Pioche, Nevada.
4. Pink to reddish argill. sh. containing interbedded ls. (0–180′).
5. Conformably overlies Lyndon Ls. in Pioche region (marine).

Coconino Sandstone — M. Permian (Leonard)

1. Northern Arizona, Southern Utah, SE Nevada.
2. Darton, 1910, U.S.G.S. Bull. 435, p. 21–27.
3. Coconino Plateau.
4. Cross-bedded white to gray ss. (50–610′).
5. Probably equivalent to De Chelly (eolian).

Crystal Pass Limestone Member (of Sultan Limestone) — U. Devonian to L. Mississippian

1. SE Nevada and SE California.
2. Hewett, 1931, U.S.G.S. Prof. Paper 162, p. 10.
3. Crystal Pass, Clark County, Nevada.
4. Thin-bedded, blue-gray ls. (150–260′).
5. Upper mbr. of Sultan Ls.; unconformably underlies Dawn Ls.; conformably overlies Valentine Ls. Mbr. of Sultan Ls.

Cutler Formation — Permian (Wolfcamp, Leonard)

1. Four Corners region, Arizona.
2. Cross and Howe, 1905, U.S.G.S. Silverton Folio No. 120.
3. Cutler Creek, 4 mi. N. of Ouray, Colorado.
4. Red arkosic ss. and cgl., sandy sh., and Ls. (0–8000′).
5. Raised to group status by Wengerd and Matheny (1958), but considered by Kirkland (1963) as a formation containing following members in central Paradox basin: Halgaito Tongue, Cedar Mesa Ss., Organ Rock Tongue, De Chelly-White Rim Ss. (continental and marine).

Dawn Limestone (of Monte Cristo Group) — L. Mississippian

1. SE Nevada.
2. Hewett, 1931, U.S.G.S. Prof. Paper 162, p. 10.
3. Dawn Mine, Goodsprings quad., Nevada.
4. Thin-bedded dark-gray ls., some dol, fossiliferous (60–400′).
5. Lowest fm. in Monte Cristo Group (marine).

De Chelly Sandstone (member of Cutler Formation) — M. Permian (Leonard)

1. SE Utah, NE Arizona, NW New Mexico.
2. Gregory, 1915, Amer. Jour. Sci., 4th Ser., v. 40, p. 102.
3. Canyon De Chelly, Apache Co., Arizona.
4. Massive, cross-bedded, light-red ss. (0–1000′).
5. Probably equivalent to White Rim Ss., raised to formation status by Wengerd and Matheny (1958) but considered member of Cutler Fm. by Kirkland (1963) (dominantly eolian).

Dunderberg Shale — U. Cambrian

1. Eastern Nevada.
2. Walcott, 1908, Smithsonian Misc. Coll., v. 53, No. 1812, p. 184.
3. Dunderberg Mine, Eureka dist., Nevada.
4. Shale.
5. Replaces Hamburg Sh. and is same as Dunderbergian Series used by Keyes (1923).

Earp Formation (of Naco Group) — Virgilian – Wolfcampian

1. SE Arizona and SW New Mexico.
2. Gilluly, Cooper, Williams, 1954, U.S.G.S. Prof. Paper 266, p. 18–23.
3. Earp Hill, central Cochise County, Arizona.
4. Red sh., ls., minor ss., and dol. (600′).
5. Uppermost unit in Naco Group (Ross, 1973); correlates with Wescogame Fm. in Grand Canyon (marine and nonmarine).

Ely Springs Dolomite — M. to U. Ordovician

1. Eastern Nevada and SE California.
2. Westgate and Knopf, 1932, U.S.G.S. Prof. Paper 171.
3. Ely Springs Range, Pioche region, Nevada.
4. Dolomite.
5. Overlies Eureka Quartzite.

Escabrosa Limestone — Mississippian

1. SE Arizona and SW New Mexico.
2. Gilluly et al., 1954, U.S.G.S. Prof. Paper 266, p. 2–13.
3. Escabrosa Ridge, Bisbee quad., Arizona.
4. Cherty crinoidal ls. (585–755′).
5. Overlies Martin Fm. or Abrigo Ls.; underlies Horquilla Ls. or Black Prince Ls. (marine).

Esplanade Sandstone — Wolfcampian

1. Grand Canyon, northern Arizona.
2. McKee, 1975, U.S.G.S. Bull. 1395–J.
3. Havasu Canyon, Grand Canyon, Arizona.
4. Sandstone with minor mdst. and slts. (400–1000′).
5. Uppermost formation of Supai Group; interfingers with Pakoon Ls. (marine).

Eureka Quartzite — M. to U. (?) Ordovician

1. Northern Nevada, western Utah, and eastern California.
2. Hague, 1883, U.S.G.S. 3rd. Ann. Rept., p. 253.
3. Long Mountain, NW of Eureka, Nevada.
4. Quartzite (100–400′).
5. Overlies Pogonip Group; underlies Ely Springs Dol.

Flour Sack Member (of Bright Angel Shale) — L. or M. Cambrian

1. NW Arizona.
2. McKee, 1945, Carnegie Inst. Washington Pub. 563, p. 14.
3. Flour Sack Rapids, Grand Canyon, Arizona.
4. Green and purple micaceous sh. and platy ls. (100′).
5. Underlies Rampart Cave Mbr. of Muav Fm. (marine).

Gateway Canyon Member (of Muav Limestone) — L. or M. Cambrian

1. NW Arizona.
2. McKee, 1945, Carnegie Inst. Wash. Pub. 563, p. 14.
3. Gateway Canyon, Grand Canyon, Arizona.
4. Thin beds of ls. and slts. (137′).
5. Thickens westward from type section; overlies Kanab Canyon Mbr.; underlies Havasu Mbr.

Goodsprings Dolomite — U. Cambrian – Devonian (?)

1. SE Nevada and SE California.
2. Hewett, 1931, U.S.G.S. Prof. Paper 162, p. 10–11.
3. Goodsprings, Clark County, Nevada.
4. Dolomite.
5. Underlies Sultan Ls.

Havasu Member (of Muav Limestone) — L. or M. Cambrian

1. NW Arizona.
2. McKee, 1945, Carnegie Inst. Wash. Pub. 563, p. 14.
3. Havasu Canyon, Grand Canyon, Arizona.

4. Mottled aphanitic ls. and fine-grained dol. (100–120').
5. Top mbr. of Muav Ls.

Hermit Shale — L. Permian (Wolfcamp)

1. Northern Arizona, southern Utah, SE Nevada.
2. Noble, 1922, U.S.G.S. Prof. Paper 131B.
3. Hermit Basin, Grand Canyon, Arizona.
4. Brick red sandy sh. and fine-grained ss. (267–317').
5. Unconformably overlies Supai Group; stratigraphic equivalent of Organ Rock Tongue; underlies Coconino Ss. (continental).

Hermosa Group — Pennsylvanian

1. Four Corners area.
2. Cross and Spencer, 1899, U.S.G.S. La Plata Folio No. 60, p. 8.
3. Hermosa Creek, San Jan Mountains, Colorado.
4. Carbonate sequence with interbedded ss., sh., and evaporite (0–6000').
5. Raised from formation to group status by Wengerd and Matheny (1958), who included in ascending order: Pinkerton Trail Fm., Paradox Fm., and Honaker Trail Fm. (marine and restricted marine and fluvial).

Honaker Trail Formation (of Hermosa Group) — U. Pennsylvanian

1. Four Corners area.
2. Wengerd and Matheny, 1958, AAPG Bull., v. 42, p. 2048–2106.
3. Honaker Trail in San Juan River Canyon, W. of Mexican Hat, Utah.
4. Interbedded ls., sh., and ss. (9–2000').
5. Previously termed upper mbr. of Hermosa Fm. (marine and fluvial).

Horquilla Limestone (of Naco Group) — M. to U. Pennsylvanian

1. SE Arizona and SW New Mexico.
2. Gilluly, Cooper, Williams, 1954, U.S.G.S. Prof. Paper 266, p. 16–18.
3. Horquilla Peak, central Cochise County, Arizona.
4. Blue-gray thin-bedded ls. with minor red shaly ls. (1000–1600').
5. Underlies Earp Fm. (marine).

Horseshoe Mesa Member (of Redwall Formation) — L. Mississippian

1. Grand Canyon, Four Corners area.
2. McKee, 1963, U.S.G.S. Prof. Paper 475–C, p. C21–C22.
3. Grandview Trail, Grand Canyon, Arizona.
4. Thin-bedded, mostly aphanitic ls., some chert beds (30–50').
5. Uppermost or "D" member in Grand Canyon (marine).

Illipah Formation — U. Mississippian– L. Pennsylvanian (?)

1. Nevada and Utah.
2. Christiansen, 1951, Utah Geol. Soc. Guidebook 6, p. 76.
3. Unknown (possibly Illipah Creek, White Pine County, Nevada).
4. Ss. and congl. tongues (?).
5. Underlies Callville Ls.; equivalent (?) to Indian Springs Fm.

Indian Springs Formation — Chesterian

1. SE Nevada.
2. Longwell and Dunbar, 1936, AAPG Bull., v. 20, p.
3. Indian Springs, NW of Las Vegas, Nevada.
4. Thin-bedded ss., sh., and ls. with local cong. (700').
5. Unconformably overlies Monte Cristo Group; conformably underlies Bird Spring Fm.

Ironside Dolomite Member (of Sultan Limestone) — M. to U. Devonian

1. SE Nevada and SE California.
2. Hewett, 1931, U.S.G.S. Prof. Paper 162, p. 10.
3. Ironside Mine, Spring Mountain Range, Nevada.
4. Dolomite (60–200').
5. Underlies Valentine Ls. Mbr. of Sultan Ls.

Jerome Member (of Martin Formation) — U. Devonian

1. Northern Arizona.
2. Stoyanow, 1930, Pan-Am. Geol., v. 53, p. 316–317.
3. Jerome, Yavapai County, Arizona.
4. Light-gray dol., locally abundant fossils (400–500').
5. Overlies Beckers Butte Mbr. of Martin Fm.; unconformably underlies Redwall Fm. (marine).

Kaibab Limestone — Permian (Leonard-Guadulupe)

1. Northern Arizona, southern Utah, SE Nevada.
2. Darton, 1910, U.S.G.S. Bull. 435.
3. Kaibab Gulch, Paria, Utah.
4. Dense, cherty, gray ls. (820').
5. Underlies Moenkopi Fm. unconformably; generally considered Leonardian, but upper part contains Guadalupian fossils near Flagstaff, Arizona (marine).

Kanab Canyon Member (of Muav Limestone) — L. or M. Cambrian

1. NW Arizona.
2. McKee, 1945, Carnegie Inst. Wash. Pub. 563, p. 14.
3. Kanab Canyon, Grand Canyon, Arizona.
4. Mottled ls. which becomes sandy sh. to SE from type section (62–143').
5. Thickness increases to west; overlies Peach Springs Mbr.; underlies Gateway Canyon Mbr. of Muav Ls.

Leadville Limestone — L. Mississippian

1. Colorado, NW New Mexico, NE Arizona, SE Utah.
2. Eldridge, 1894, U.S.G.S. Anthracite – Crested Butte Folio No. 9.
3. Leadville mining district, Colorado.
4. Massive ls. and dol., with interbeds of sh. and chert. (0–200').
5. Overlies Ouray Fm.; unconformably underlies Molas Fm.; equivalent to Redwall Ls. (marine).

Lynch Dolomite — U. Cambrian

1. Central and east-central Utah.
2. Gilluly, 1932, U.S.G.S. Prof. Paper 173.
3. Lynch Ridge, N of Ophir, Utah.
4. Light to dark-gray dol. (0–1400').
5. Uppermost Cambrian fm. in east-central Utah and Paradox Basin (marine).

Lyndon Limestone — M. Cambrian

1. Eastern Nevada (Pioche region).
2. Westgate and Knopf, 1927, Am. Inst. Min. Met. Engrs. Trans., No. 1647, and U.S.G.S. Prof. Paper 171.
3. Lyndon Gulch, Nevada.
4. Light-gray to white crystalline ls. (200').
5. Conformably underlies Chisholm Sh.; conformably overlies Pioche Sh.

Manakacha Formation — L. Pennsylvanian

1. Grand Canyon, Arizona.
2. McKee, 1975, U.S.G.S. Bull. 1395–J.
3. Havasu Canyon, Grand Canyon.
4. Sandy ls. with minor mdst. (+300′).
5. Rests unconformably on Redwall Ls.; underlies Wescogame Fm. of Supai Group (marine).

Martin Formation — M. to U. Devonian

1. Central and eastern Arizona.
2. Ransome, 1904, U.S.G.S. Prof. Paper 21.
3. Mt. Martin, SE Arizona.
4. Gray fine-grained ls. and dol. (200–300′).
5. Jerome Mbr. present in Verde Valley region, Arizona; unconformably underlies Redwall Fm. (marine).

Meriwitica Tongue (of Bright Angel Shale) — L. or M. Cambrian

1. NW Arizona.
2. McKee, 1945, Carnegie Inst. Washington Pub. 563, p. 14.
3. Meriwitica Canyon, Grand Canyon, Arizona.
4. Rusty brown dol. (16′).
5. Underlies Flour Sack Mbr. of Bright Angel Sh.

Molas Formation — L. Pennsylvanian

1. Four Corners area.
2. Cross and Howe, 1905, U.S.G.S. Silverton Folio No. 120, and GSA Bull., v. 16, p. 470–496.
3. Molas Lake, near Silverton, Colorado.
4. Red variegated sh., thin ls. and ss. (0–200′).
5. Represents residual and reworked soil mantle on Leadville Fm.; locally includes three mbrs. (continental and marine).

Monte Cristo Group — L. to U. Mississippian

1. SE Nevada and SE California.
2. Hewett, 1931, U.S.G.S. Prof. Paper 162, p. 9–10.
3. Monte Cristo Mine, Goodsprings quad., Nevada.
4. Ls. and dol. (350–1600′).
5. Composed of 5 fm. in ascending order: Dawn Ls., Anchor Ls., Bullion Dol., Arrowhead Ls., Yellowpine Ls.

Mooney Falls Member (of Redwall Formation) — L. Mississippian

1. Grand Canyon, Four Corners area.
2. McKee, 1963, U.S.G.S. Prof. Paper 475–C, p. 21–22.
3. Havasu Canyon, Grand Canyon, Arizona.
4. Thick-bedded, massive, cliff-forming ls., aphanitic to crinoidal (50–325′+).
5. Member "C"; underlies Horseshoe Mesa Mbr. (marine).

Mountain Springs Formation — M. Ordovician to Devonian

1. SE Nevada.
2. Gans, 1974, Geol. Soc. Am. Bull., v. 85, p. 196–198.
3. Hill 6,646, Shenandoah Peak quad., Nevada.
4. Coarsely crystalline dol., 116 m.
5. Dolomitized marine ls.; replaces "Ely Springs Dolomite."

Muav Limestone (of Tonto Group) — M. Cambrian

1. Northern Arizona and SE Utah.
2. Noble, 1914, U.S.G.S. Bull. 549.
3. Muav Canyon, Grand Canyon, Arizona.
4. Blue-gray, thin-bedded, mottled buff sh. and ls. (450–475′).
5. Equivalent to Maxfield Ls. of central Utah, contains seven mbrs., in ascending order: Rampart Cave, Sanup Plateau, Spencer Canyon, Peach Springs, Kanab Canyon, Gateway Canyon, and Havasu Mbrs. (marine).

Muddy Peak Limestone — M. Devonian

1. SE Nevada, SW Utah, and NW Arizona.
2. Longwell, 1921, Am. Jour. Sci., 5th, v. 1, p. 46–53.
3. Muddy Peak, Muddy Mountains, Nevada.
4. Limestone (550′).
5. Equivalent to Sultan Ls.

Naco Formation — Pennsylvanian

1. Eastern Arizona.
2. Ransome, 1904, U.S.G.S. Prof. Paper 21.
3. Naco Hills, Bisbee quad., Arizona.
4. Crinoidal and cherty ls. interbedded with red sh. (2000′).
5. Indistinguishable from underlying Molas Fm. in Black Mesa Basin (continental to marine).

Nopah Formation — U. Cambrian

1. SE California and western Nevada.
2. Hazzard, 1937, Calif. Jour. Mines and Geol., v. 33, p. 276.
3. Nopah Mts., Inyo County, California.
4. Light-gray dol. (1600–1740′).
5. Conformably underlies Pogonip Fm.; overlies Bonanza King Fm.

Pakoon Limestone — L. Permian (Wolfcamp)

1. NW Arizona and SW Utah.
2. McNair, 1951, AAPG Bull., v. 35, p. 515, 524–25.
3. Pakoon Ridge, NW of Pakoon Springs, Mohave County, Arizona.
4. Gray ls. and dol. (688′).
5. Overlies Callville Ls.; underlies Queantoweap Ss.; probably equivalent to Elephant Canyon Fm. of northwestern Paradox Basin.

Papoose Lake Member (of Bonanza King Formation) — M. Cambrian

1. SE Nevada.
2. Barnes and Palmer, 1961, U.S.G.S. Prof. Paper 424–C, p. C101.
3. Papoose Lake Playa, Banded Mountain, Nye County, Nevada.
4. Ls. and dol. with minor slts. (2700′).
5. Underlies Banded Mt. Mbr. of Bonanza King Fm.

Paradox Formation (of Hermosa Group) — Pennsylvanian

1. Four Corners area (Paradox Basin).
2. Baker, 1933, U.S.G.S. Bull. 841.
3. Paradox Valley, Montrose County, Colorado.
4. Cyclic deposits of salt, anhydrite, black sh., dol. (0–5000′?)
5. Evaporite facies of Hermosa Group. (restricted marine).

Peach Springs Member (of Muav Limestone) — L. or M. Cambrian

1. NW Arizona.
2. McKee, 1945, Carnegie Inst. Wash. Pub. 563, p. 14.
3. Peach Springs Wash, western Grand Canyon, Arizona.
4. Thin-bedded gray ls. and slst. (42–90′).
5. Underlies Kanab Canyon Mbr.; overlies Spencer Canyon Mbr.

Pinkerton Trail Formation (of Hermosa Group) — Pennsylvanian

1. Four Corners region (Paradox basin).
2. Wengerd and Strickland, 1954, AAPG Bull., v. 38, p. 2157–99.

3. Pinkerton Trail, N of Durango, Colorado.
4. Gray ls., sh.,slts., and ss. (0–300′).
5. Lowest unit of Hermosa Group (marine).

Pioche Shale — L. Cambrian

1. Western Utah and eastern Nevada.
2. Walcott, 1908, Smithsonian Misc. Coll., v. 53, no. 1804, p. 9–12.
3. SE of Pioche, Nevada.
4. Sh. with interbedded ls., contains *Olenellus* fauna (125–1120′).
5. Conformably overlies Prospect Mountain Quartzite, equivalent to Bright Angel Sh. (marine).

Pogonip Formation — L. to M. Ordovician

1. Eastern and southern Nevada, SE California, and western Utah.
2. King, 1876, U.S. Geol. Expl. 40th Par. Atlas, Map IV.
3. Pogonip Ridge, 30 miles southeast of Eureka, Nevada.
4. Interbedded sh. and ls.
5. Conformably overlies Nopah Fm.; underlies Eureka Quartzite.

Prospect Mountain Quartzite — L. Cambrian

1. Eastern Nevada, eastern California, and western Utah.
2. Hague, 1883, U.S.G.S. 3rd Ann. Rept., p. 253–259.
3. Prospect Peak, Nevada.
4. Quartzite (1000–3000′).
5. Underlies Pioche Sh.; equivalent to Tapeats Ss. (marine).

Queantoweap Sandstone — L. Permian (Wolfcamp)

1. NW Arizona and SW Utah.
2. McNair, 1951, AAPG Bull., v. 35, p. 515, 524–25.
3. East side of Queantoweap Valley, Mohave County, Arizona.
4. Massive pink and gray cross-bedded ss. (393–645′).
5. Overlies Pakoon Ls.; transitionally underlies Hermit Sh.; probably equivalent to upper Supai–Hermit Sh. and Cedar Mesa Ss.–Organ Rock Sh. (shallow marine and eolian?).

Rampart Cave Member (of Muav Limestone) — L. or M. Cambrian

1. Grand Canyon, Arizona.
2. McKee, 1945, Carnegie Inst. Wash. Pub. 563, p. 14.
3. Rampart Cave, Grand Canyon, Arizona.
4. Gray aphanitic ls. (160′).
5. Basal unit of Muav thins eastward; overlies Flour Sack Mbr. of Bright Angel Sh.

Redwall Limestone — Mississippian

1. Northern Arizona and south-central Utah.
2. Gilbert, 1875, U.S.G. & G.S., W. 100th Meridian, v. 3, p. 162.
3. Grand Canyon, Arizona.
4. Ls. and dol., light colored, cherty (500–700′).
5. Unconformably overlies Temple Butte; unconformably underlies Supai Group. McKee divided unit into 4 members, in ascending order: Whitmore Wash., Thunder Springs, Mooney Falls, and Horseshoe Mesa Mbrs. (marine).

Sanup Plateau Member (of Muav Limestone) — L. or M. Cambrian

1. NW Arizona.
2. McKee, 1945, Carnegie Inst. Wash. Pub. 563, p. 14.
3. Sanup Plateau, western Grand Canyon, Arizona.
4. Dark-gray aphanitic ls. with bands of slst. becomes algal ls. to east and south (25-30′).
5. Underlies Spencer Canyon Mbr.; overlies Rampart Cave Mbr.

Spencer Canyon Member (of Muav Limestone) — L. or M. Cambrian

1. NW Arizona.
2. McKee, 1945, Carnegie Inst. Wash. Pub. 563, p. 14.
3. Western Grand Canyon, Arizona.
4. Gray aphanitic ls., slst., and minor sh. (40′).
5. Overlies Sanup Plateau Mbr.; underlies Peach Springs Mbr.

Spring Mountain Formation — Wolfcamp–Leonard

1. Western Nevada.
2. Bissell, 1960, AAPG Bull., v. 44, no. 8, p. 1427.
3. Spring Mountain, Nevada.
4. Limestone.
5. Overlies Bird Spring Fm.; underlies Kaibab Ls.; equivalent to Permian part of Bird Spring Fm.

Sultan Limestone — M. or U. Devonian

1. SE Nevada.
2. Hewett, 1931, U.S.G.S. Prof. Paper 162, p. 10.
3. Sultan Mine, Goodsprings quad., Nevada.
4. Upper part ls., dol. in lower part (230-760′).
5. Contains three mbrs., in ascending order: Ironside Dol., Valentine Ls., and Crystal Pass Ls. Mbrs., overlain by Monte Cristo Group and underlain by Goodsprings Dol.

Supai Group — L. Pennsylvanian to L. Permian

1. Grand Canyon, northern Arizona, southern Utah, SE Nevada.
2. McKee, 1975, U.S.G.S. Bull. 1395-J, 11 p.
3. Havasu Canyon, Grand Canyon, Arizona.
4. Red ss., sh., and gray ls. (500-1000′).
5. Partially equivalent to Cutler Fm.; includes in ascending order: Watahomigi Fm., Manakacha Fm., Wescogame Fm., Esplanade Ss. (marine).

Tapeats Sandstone (of Tonto Group) — L. and M. Cambrian

1. Northern Arizona, Grand Canyon and SE Utah, west of Monument Uplift and La Sal Mts.
2. Noble, 1914, U.S.G.S. Bull. 549.
3. Tapeats Creek, Grand Canyon, Arizona.
4. Cross-bedded ss. and qtzite, with cgl. lenses (0-285′+).
5. Rests unconformably on Precambrian (transgressive marine).

Temple Butte Formation — U. Devonian

1. Northern Arizona.
2. Walcott, 1889, GSA Bull., v. 1, p. 50.
3. Temple Butte, Grand Canyon National Park, Arizona.
4. Purple and cream ls. and ss. (0-100′).
5. Applied in Grand Canyon area; approximate equivalent of Elbert Fm. of Four Corners area (marine).

Thunder Springs Member (of Redwall Formation) — L. Mississippian

1. Grand Canyon, Four Corners area, Arizona.
2. McKee, 1963, U.S.G.S. Prof. Paper 475-C, p. C21-C22.
3. Thunder River Trail, Grand Canyon, Arizona.
4. Alternate beds of chert and carbonate, forming banded cliffs (25-175′+).
5. Member "B"; underlies Mooney Falls Mbr. (marine).

Tincanebits Tongue (of Bright Angel Shale) — L. to M. Cambrian

1. Grand Canyon, Arizona.
2. McKee, 1945, Carnegie Inst. Wash. Pub. 563, p. 14.
3. Western Grand Canyon, Arizona.
4. Rusty brown dol. tongue (8-30′).
5. May constitute a member of the Muav.

Tonto Group — L. and M. Cambrian

1. Grand Canyon and northern Arizona.
2. Gilbert, 1874, Phil. Soc. Wash. Bull., v. 1, p. 109.
3. Tonto Creek, tributary of Salt River, Arizona.
4. Ss. and quartzite, sh., ls. (800′±).
5. Contains following formations in ascending order: Tapeats Ss., Bright Angel Sh., Muav Ls. (marine).

Toroweap Formation — M. Permian (Leonard)

1. Nevada, SW Utah, NW Arizona.
2. McKee, 1937, Carnegie Inst. Wash. Pub. No. 492.
3. East wall of Toroweap Valley, tributary of Grand Canyon, Arizona.
4. Red beds, ss., gypsum, and massive ls. (290′).
5. Underlies Kaibab; overlies Coconino in western Arizona (continental to marine).

Valentine Limestone Member (of Sultan Limestone) — M. to U. Devonian

1. SE Nevada and SE California.
2. Hewett, 1931, U.S.G.S. Prof. Paper 162, p. 10.
3. Valentine Mine, Clark County, Nevada.
4. Limestone (75-380′).
5. Overlies Ironside Mbr.; underlies Crystal Pass Limestone Mbr. (both of Sultan Ls.).

Watahomigi Formation — L. Pennsylvanian

1. Grand Canyon, Arizona.
2. McKee, 1975, U.S.G.S. Bull. 1395–J.
3. Havasu Canyon, Grand Canyon.
4. Ls. and mdst. (200–250′).
5. Lowest fm. of Supai Group (marine).

Wescogame Formation — Virgilian

1. Grand Canyon, Arizona.
2. McKee, 1975, U.S.G.S. Bull. 1395–J.
3. Havasu Canyon, Grand Canyon.
4. Ss. and sandy ls. (250′).
5. Rests unconformably on Manakacha Fm.; unconformably underlies Esplanade Ss. (marine).

Whitmore Wash Member (of Redwall Formation) — L. Mississippian

1. Grand Canyon, Four Corners area, Arizona.
2. McKee, 1963, U.S.G.S. Prof. Paper 475–C, p. 21–22.
3. Whitmore Wash north of Colorado River, Grand Canyon.
4. Basal thick-bedded unit, dol. in Four Corners and eastern Grand Canyon (50–200′).
5. Member "A"; underlies Thunder Springs Mbr. (marine).

Yellowpine Limestone (of Monte Cristo Group) — U. Mississippian

1. SE Nevada.
2. Hewett, 1931, U.S.G.S. Prof. Paper 162, p. 9.
3. Yellow Pine Mine, Nevada.
4. Dark-gray ls. and minor dol. (60–120′).
5. Top fm. of Monte Cristo Group; overlies Arrowhead Ls.